W9-AOB-355

BIOSAFETY IN THE LABORATORY

Prudent Practices for the Handling and Disposal of Infectious Materials

Committee on Hazardous Biological Substances in the Laboratory
Board on Chemical Sciences and Technology
Commission on Physical Sciences, Mathematics, and Resources
National Research Council

NATIONAL ACADEMY PRESS
Washington, D.C. 1989

NATIONAL ACADEMY PRESS • 2101 Constitution Avenue, NW • Washington, DC 20418

NOTICE: The project that is the subject of this report was approved by the Governing Board of the National Academy of Sciences, the National Academy of Engineering, and the Institute of Medicine. The members of the committee responsible for the report were chosen for their special competences and with regard for appropriate balance.

This report has been reviewed by a group other than the authors according to procedures approved by a Report Review Committee consisting of members of the National Academy of Sciences, the National Academy of Engineering, and the Institute of Medicine.

The National Academy of Sciences is a private, nonprofit, self-perpetuating society of distinguished scholars engaged in scientific and engineering research, dedicated to the furtherance of science and technology and to their use for the general welfare. Upon the authority of the charter granted to it by the Congress in 1863, the Academy has a mandate that requires it to advise the federal government on scientific and technical matters. Dr. Frank Press is the president of the National Academy of Sciences.

The National Academy of Engineering was established in 1964, under the charter of the National Academy of Sciences, as a parallel organization of outstanding engineers. It is autonomous in its administration and in the selection of its members, sharing with the National Academy of Sciences the responsibility for advising the federal government. The National Academy of Engineering also sponsors engineering programs aimed at meeting national needs, encourages education and research, and recognizes the superior achievements of engineers. Dr. Robert M. White is president of the National Academy of Engineering.

The Institute of Medicine was established in 1970 by the National Academy of Sciences to secure the services of eminent members of appropriate professions in the examination of policy matters pertaining to the health of the public. The Institute acts under the responsibility given to the National Academy of Sciences by its congressional charter to be an adviser to the federal government and, upon its own initiative, to identify issues of medical care, research, and education. Dr. Samuel O. Thier is president of the Institute of Medicine.

The National Research Council was organized by the National Academy of Sciences in 1916 to associate the broad community of science and technology with the Academy's purpose of furthering knowledge and advising the federal government. Functioning in accordance with general policies determined by the Academy, the Council has become the principal operating agency of both the National Academy of Sciences and the National Academy of Engineering in providing services to the government, the public, and the scientific and engineering communities. The Council is administered jointly by both Academies and the Institute of Medicine. Dr. Frank Press and Dr. Robert M. White are chairman and vice chairman, respectively, of the National Research Council.

Support for this project was provided by the U.S. Department of Agriculture and National Institutes of Health under contract no. 59-32U4-5-33; the U.S. Department of the Army under contract no. DAMD17-86-G-6022; the U.S. Department of Energy under grant no. DE-FG05-85ER13457; the U.S. Department of Labor under purchase order no. B9F56292; and the National Science Foundation under grant no. DMB-8611553. The project has also been supported by funds from the National Research Council Fund, a pool of private, discretionary, nonfederal funds that is used to support a program of Academy-initiated studies of national issues in which science and technology figure significantly. The NRC Fund consists of contributions from a consortium of private foundations including the Carnegie Corporation of New York, the Charles E. Culpeper Foundation, the William and Flora Hewlett Foundation, the John D. and Catherine T. MacArthur Foundation, the Andrew W. Mellon Foundation, the Rockefeller Foundation, and the Alfred P. Sloan Foundation, as well as the Academy Industry Program, which seeks annual contributions from companies that are concerned with the health of U.S. science and technology and with public policy issues with technological content.

Library of Congress Cataloging-in-Publication Data

National Research Council (U.S.). Committee on Hazardous Biological Substances in the Laboratory.
Biosafety in the laboratory: prudent practices for the handling and disposal of infectious materials / Committee on Hazardous Biological Substances in the Laboratory, Board on Chemical Sciences and Technology, Commission on Physical Sciences, Mathematics, and Resources, National Research Council.
p. cm.
Includes bibliographical references.
ISBN 0-309-03975-4
1. Medical laboratories—Safety measures—Congresses. 2. Medical laboratories—Waste disposal—Safety measures—Congresses.
I. Title.
[DNLM: 1. Containment of Biohazards—standards—United States. 2. Laboratories—standards—United States. 3. Laboratory Infection—prevention & control—United States. QY 23 N2775b]
R860.N37 1989
616.9'0072—dc20 89-13004
DNLM/DLC for Library of Congress CIP

Printed in the United States of America

Committee on Hazardous Biological Substances in the Laboratory

---

Committee (Authors)

EDWARD A. ADELBERG (*Chairman*), Yale University
ROBERT AUSTRIAN, University of Pennsylvania School of Medicine
HOWARD L. BACHRACH, Agricultural Research Service, U.S. Department of Agriculture
W. EMMETT BARKLEY, National Institutes of Health
J. PAUL BURNET, Lilly Research Laboratories
DIANE O. FLEMING, Sterling Research Group
ROY L. FUCHS, Monsanto Company
HAROLD S. GINSBERG, Columbia University
ROSE GOLDMAN, Cambridge Hospital, Harvard University School of Medicine
JAMES M. HUGHES, Centers for Disease Control
WILLIAM G. MIKELL, E.I. du Pont de Nemours & Company, Inc.
JOHN H. RICHARDSON, Emory University
JEROME P. SCHMIDT, USAF School of Aerospace Medicine
JAMES W. SMITH, Indiana University School of Medicine
THOMAS E. WALTON, Agricultural Research Service, U.S. Department of Agriculture

NRC Staff

WILLIAM SPINDEL, *Study Director* (July 1985 to January 1987)
ROBERT M. SIMON, *Study Director* (January 1987 to July 1989)
BENNETT L. ELISBERG, *Consultant*
MONALISA BRUCE, *Administrative Secretary*
SANDRA NOLTE, *Administrative Secretary*

Contributors

MICHELLE EVANS, National Institutes of Health
GREGG J. HUNT, Agricultural Research Service, U.S. Department of Agriculture
JOHN KEENE, Abbott Laboratories
GEORGE P. KUBICA, Centers for Disease Control
ROBERT W. McKINNEY, National Institutes of Health
JONATHAN RICHMOND, National Institutes of Health
HARVEY W. ROGERS, National Institutes of Health
CLARENCE STYRON, Monsanto Company
JERRY TULIS, University of North Carolina
DONALD VESLEY, University of Minnesota

Board on Chemical Sciences and Technology

Commission on Physical Sciences, Mathematics, and Resources

Preface

In 1981 and 1983, the National Research Council published two reports on chemical safety in the laboratory: *Prudent Practices for Handling Hazardous Chemicals in Laboratories* and *Prudent Practices for Disposal of Chemicals from Laboratories* [93,94]. In November 1983, a planning committee of the National Research Council was convened under the chairmanship of Thomas Weller to consider the need for a document that would deal in a similar way with biological safety in the laboratory. The committee concluded that such a document would be timely and recommended the formation of a working committee to produce it.

The Committee on Hazardous Biological Substances in the Laboratory was organized in the fall of 1985. It was presented with a broad charge, namely, to prepare a report dealing with the following aspects of hazardous biological materials:

1. Definition of laboratory safety problems with hazardous biological materials.
2. Guidelines for physical facilities, equipment, and work practices.
3. Procedures for identifying hazards and establishing conditions for any operation involving hazardous biological materials.
4. Guidelines for waste disposal, including incinerating, venting, and discharging to sewer systems.
5. Guidelines for all aspects of an effective safety program including medical surveillance, compliance with regulations, and recordkeeping.
6. A plan for obtaining consensus on and implementation of the guidelines.

The committee first met in January 1986 and decided to restrict the scope of the report to the safe handling and disposal of agents hazardous to humans; strict animal pathogens and strict plant pathogens were considered to be of interest to different, specialized audiences, and better dealt with in other publications. It was also decided to deal only briefly with such hazardous biological products as toxins and immunoactive substances.

During the period in which this report was being planned and written, a number of excellent books appeared dealing with various aspects of biosafety in the laboratory (see, for example, references 4, 83, and 149). Although the present report overlaps many sections of these books, the committee felt that the need still existed for a consensus, peer-reviewed document, produced under the imprimatur of the NRC, that could serve as a general set of guidelines for the safe handling and disposal of infectious materials in the laboratory. This book represents the committee's efforts to produce such a document, with the able support of the National Research Council's staff: in particular, we wish to thank William Spindel, Robert M. Simon, and Bennett L. Elisberg for their expert assistance. We also wish to acknowledge the contributions of the 30 or more reviewers, representing every type of microbiological laboratory, whose thoughtful and constructive comments formed the basis of many changes in the final draft of this book.

EDWARD A. ADELBERG, *Chairman*
Committee on Hazardous Biological
Substances in the Laboratory

Contents

1
INTRODUCTION, OVERVIEW, AND RECOMMENDATIONS / 1

2
DESCRIPTIVE EPIDEMIOLOGY OF OCCUPATIONAL INFECTIONS OF LABORATORY WORKERS / 8

3
SAFE HANDLING OF INFECTIOUS AGENTS / 13

5
SAFETY MANAGEMENT / 46

BIOSAFETY
IN THE LABORATORY

1

Introduction, Overview, and Recommendations

A. INTRODUCTION

This book is about the safe handling and disposal of hazardous biological materials in the laboratory. These materials consist of infectious agents, per se, as well as substances actually or potentially contaminated with them.

A large number of laboratory workers handle such materials as part of their daily routine. The number has been estimated to be about 500,000 in the United States, but that number is probably a gross underestimate. The persons at risk are primarily the laboratory workers themselves, but the risks may extend to others: students, custodial and maintenance workers who must enter laboratories, handlers of shipped materials, sanitation workers, and all who work in or pass through building areas adjacent to the laboratory.

For the purposes of this book, the term "risk" refers to the probability of acquiring an occupational infection, rather than to the severity of the resulting disease. Such risk is actually much lower than it is popularly perceived to be: the great majority of organisms handled in the laboratory are either not known to be hazardous or are of minimal potential hazard to laboratory personnel; in any case the risk of exposure can be reduced to a very low level by the use of the simple, prudent practices described in this book. Furthermore, there is little or no risk to the community at large: disease outbreaks in the United States attributable to the escape of infectious agents, either from laboratories or from waste disposal sites, have been extremely rare.

The general concepts set forth in this book apply to many types of laboratories: academic, industrial, and governmental research laboratories; hospital, physicians', veterinarians', and dentists' laboratories; teaching laboratories; blood banks; and analytical laboratories that handle potentially infectious materials (e.g., clinical, diagnostic, and food laboratories). We have included a section on biosafety in large-scale production facilities, because laboratory workers are often involved in the scale-up of benchtop operations to the pilot plant level.

We have restricted ourselves to agents infectious to humans: strictly animal and plant pathogens are not addressed, although many of the practices recommended here are useful in the prevention of their spread. We deal only briefly with the hazards associated with biological products (e.g., toxins and immunoactive materials), and we have chosen not to include the subject of recombinant DNA, given its extensive coverage elsewhere [134,136] and the fact that the hazard presented by an organism is not related to the use of recombinant technology in its production but rather to the relative pathogenicity of the donor organism, the nature of the vector, and the hardiness of the recipient (host) organism. We do not deal in this book with problems specific to hospital wards, to nosocomial infections, or to environmental situations such as the presence of legionellas in water tanks.

This book is designed to serve as an introductory guide to biological safety in the laboratory; the principles and practices described, however, are general ones, and the readers must decide how best to apply them in their own circumstances. Specific applications will vary with such factors as the design of the local facility and equipment, the procedures in use, the nature of the potential exposure, and the workers' susceptibility.

Our text is addressed to all who are responsible

for the safety of others, including the chief executive officer of the institution, departmental chairpersons and managers, project directors, and laboratory supervisors; it is also addressed to the individual laboratory workers, who share the responsibility for their own safety as well as the safety of those around them.

Although we have tried to be comprehensive in our treatment of the general principles of biological safety in the laboratory, we have dealt only briefly with specific practices that are fully treated elsewhere. A list of pertinent references is provided, and the reader is urged to consult them for a more complete treatment of the subject. In particular, laboratory workers who come into contact with human blood, body fluids, or tissue should pay special attention to the practices described in the proposed guidelines on "Protection of Laboratory Workers from Infectious Disease Transmitted by Blood and Tissues," published by the National Committee for Clinical Laboratory Standards [90], and the Centers for Disease Control publication, "Recommendations for Prevention of HIV Transmission in Health-Care Workers" [34].

Finally, we call the reader's attention to Appendix A, which reprints in its entirety the Centers for Disease Control (CDC)/National Institutes of Health (NIH) publication entitled *Biosafety in Microbiological and Biomedical Laboratories,* published in 1984; Appendix B, the 1988 "Agent Summary Statement for Human Immunodeficiency Viruses (HIVs)," the etiologic agent of AIDS; and Appendix C, "Recommendations for Prevention of HIV Transmission in Health-Care Settings" from the *Morbidity and Mortality Weekly Report* of the CDC. Together, these documents provide guidance for the handling of most infectious agents that pose significant risks in the laboratory.

B. OVERVIEW

The remainder of this book is divided into four chapters: Chapter 2 deals with the epidemiology of laboratory-acquired infections; Chapter 3 with the safe handling of infectious materials; Chapter 4 with the safe disposal of infectious materials; and, Chapter 5 with safety management. Following is a brief overview of these chapters.

Chapter 2. Descriptive Epidemiology of Occupational Infections of Laboratory Workers

To determine the rate of occupationally related infections among laboratory workers, it is necessary to know both the number of actual infections over a given period of time and the number of persons who are at risk. As discussed in this chapter, neither figure is known; however, the data presented here give some qualitative appreciation of risk and group pathogens into high-risk and low-risk categories. Thus, from the beginning of the twentieth century until the present, the "top five" (highest-risk) organisms handled in laboratories have been the agents of brucellosis, Q fever, typhoid fever, viral hepatitis, and tuberculosis; organisms in the category of lowest risk have included rabies virus, Creutzfeldt-Jakob agent, *Vibrio cholerae, Clostridium tetani, C. botulinum,* and HIV. These rank orders are changing, however, with changes in the prevalence of the microorganisms in the general population, and in the frequency with which they are handled in the laboratory.

Chapter 3. Safe Handling of Infectious Agents

The material in this chapter falls into two categories. The first category, comprising Sections A through E, deals with the biological materials that may be hazardous:

- **pathogenic microorganisms;**
- **organisms posing special risks;**
- **vertebrate animals and insects;**
- **cell cultures; and**
- **necropsy and surgical specimens.**

In each case, there is a discussion of the special precautions that should be taken in handling the organism or the infectious material.

The second category, comprising Sections F through K, deals with the general procedures and equipment that make it possible to handle biohazardous materials safely. Section F, "Good Laboratory Practices," lists seven basic rules of biosafety; these are reprinted in the set of recommendations at the end of this chapter (Section C.3).

The same good laboratory practices are recommended for persons working with plant-specific pathogens, animal-specific pathogens, and other viable microorganisms not associated with human disease.

The remaining sections deal with the following topics:

- **transportation and shipment of biomedical materials,**
- **labeling of specimens,**
- **prevention of aerosol and droplet generation,**
- **containment equipment,**
- **personnel protective equipment,**
- **biosafety in large-scale production, and**
- **small-volume clinical laboratories.**

Once again it must be emphasized that the principles and practices described in this book are general ones, and the readers must decide how best to apply them under their own special circumstances.

Chapter 4. Safe Disposal of Infectious Laboratory Waste

In the United States, biological laboratory waste presents an occupational, rather than a public health, hazard. For the reasons discussed in this chapter, outbreaks of infectious disease attributable to such waste have not occurred or are extremely rare as a result of highly effective sanitation measures. These measures include physical barriers in the form of well-constructed drains, sewers, and refuse containers, along with properly constructed and operated sewage treatment plants, sanitary landfills, and municipal incinerators. Thus the public's health is protected even if infectious waste is introduced untreated into a municipal sewage treatment facility, a sanitary landfill, or a solid waste incinerator.

The potential for exposure to infectious agents does exist, however, for workers who generate, handle, and process biological laboratory waste. These workers can be protected effectively from exposure by a number of simple procedures, of which the principal ones are as follows:

- **segregation of infectious from noninfectious waste;**
- **on-site treatment, including chemical decontamination, use of the steam autoclave or incinerator, and appropriate packaging for transport; and**
- **personal protection, in the form of protective clothing, gloves, and handwashing.**

Mixed waste, which contains infectious agents and radioactive or chemically hazardous materials, requires special processing.

The primary responsibility for the safe handling and disposal of infectious waste lies with the laboratory that generates the waste, but waste haulers and managers of treatment facilities and sanitary landfills also share in the responsibility.

Chapter 5. Safety Management

This chapter outlines the administrative responsibilities associated with biosafety in the laboratory and recommends some general practices for dealing with them. As described here, an effective safety management program includes the following:

- **clear goals,**
- **well-defined responsibilities,**
- **mandatory safety rules,**
- **written safety plans,**
- **safety committees,**
- **effective safety communications,**
- **emergency preparedness, and**
- **auditing of laboratory operations.**

Also, as discussed in this chapter, every institution or laboratory should have a biosafety manual that addresses the following topics:

- **policy and goals,**
- **safety organization,**
- **medical programs,**
- **laboratory procedures for labeling and handling specimens, preventing aerosol and droplet generation, properly using needles, discarding materials, steam autoclaving and disinfecting, cleaning-up of spills, using and maintaining safety cabinets, controlling insects and other pests, and working with animals,**
- **safety equipment,**
- **waste disposal, and**
- **emergencies.**

The above outlines for a safety plan and a biosafety manual can serve as a rough checklist for those who are responsible for day-to-day safety in the laboratory.

Chapter 5 also deals with the proper design, operation and maintenance of facilities; safety training; risk assessment; record keeping; medical surveillance; and regulation and accreditation.

C. RECOMMENDATIONS

On the basis of the considerations discussed in this report, the committee recommends the following actions; they are presented in the order in which they appear in the text:

1. Immunization

Employees handling clinical specimens or infectious agents should be immunized with the vaccines required for admission to elementary school, or have documented immunity. Personnel working with blood, serum, or other body fluids should be immunized against hepatitis B. Immunizations with other available vaccines, including experimental products, should be considered on an individual basis. (See Chapter 3, Sections B and E, and Chapter 5, Section D.)

2. Serum Bank

The establishment of a serum bank should be considered for employees, depending upon the job situation and management policy. Specimens should be collected at the time of employment and others taken periodically thereafter. (See Chapter 3, Section B, and Chapter 5, Section D.)

3. Avoiding Exposure to Infectious Agents

The following seven laboratory practices should be observed at all times.

- **Do not mouth pipette (Figure 1.1).**
- **Manipulate infectious fluids carefully to avoid spills and the production of aerosols and droplets (Figure 1.2).**
- **Restrict the use of needles and syringes to those procedures for which there are no alterna-**

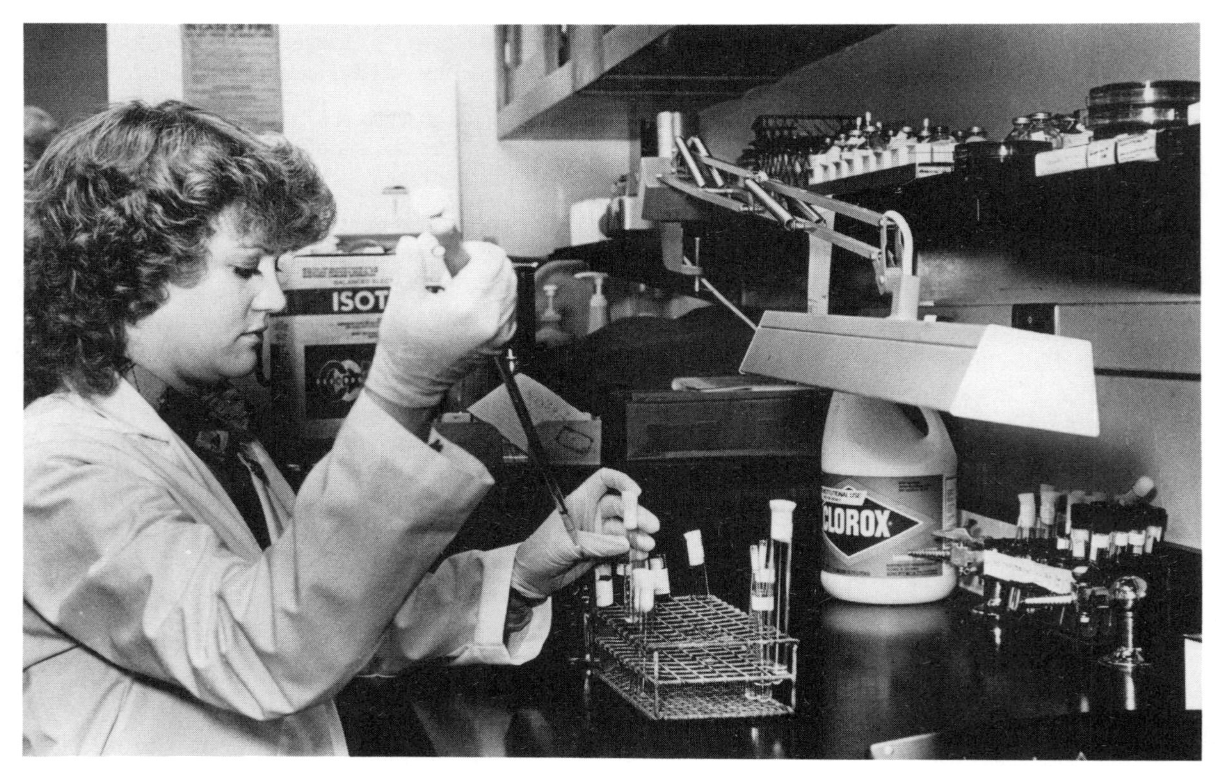

FIGURE 1.1 Do not mouth pipette. Regrettably, many laboratory workers have been taught to pipette by mouth, a practice that has led to a significant number of laboratory-acquired infections. These individuals should be required to give up the old practice and to learn to use the pipetting aids that are now available for any application. Any device requiring mouth suction should be considered unsafe and inappropriate for use in the biological laboratory. Courtesy, John H. Richardson.

tives; use needles, syringes, and other "sharps" carefully to avoid self-inoculation; and dispose of "sharps" in leak- and puncture-resistant containers.

- **Use protective laboratory coats and gloves.**
- **Wash hands following all laboratory activities, following the removal of gloves, and immediately following contact with infectious materials (Figure 1.3).**
- **Decontaminate work surfaces before and after use, and immediately after spills (Figure 1.4).**
- **Do not eat, drink, store foods, or smoke in the laboratory.**

In working with specific etiologic agents, the NIH/CDC guidelines reprinted in Appendix A should be followed. (See Chapter 3, Section F.)

4. Transportation and Shipment of Specimens

All shipments of biological materials, cell cultures, and infectious agents should be made in accordance with the applicable regulations of the U.S. Public Health Service, the U.S. Department of Transportation, the U.S. Department of Agriculture, and the U.S. Postal Service. (See Chapter 3, Section G.)

FIGURE 1.2 Manipulate infectious fluids carefully to avoid spills and the production of aerosols and droplets. This photomicrograph shows the copious production of aerosols and droplets when the last drop in a pipette is blown out. Enough material can be aerosolized by such practices to create an infectious dose of some agents. Courtesy, National Institutes of Health.

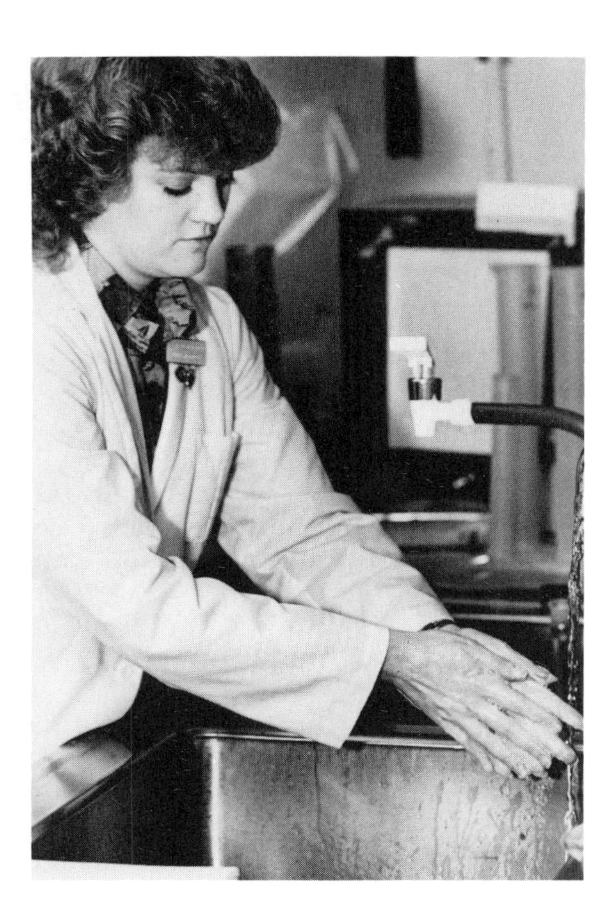

FIGURE 1.3 Wash hands following all laboratory activities, following the removal of gloves, and immediately following contact with infectious agents. Courtesy, John H. Richardson.

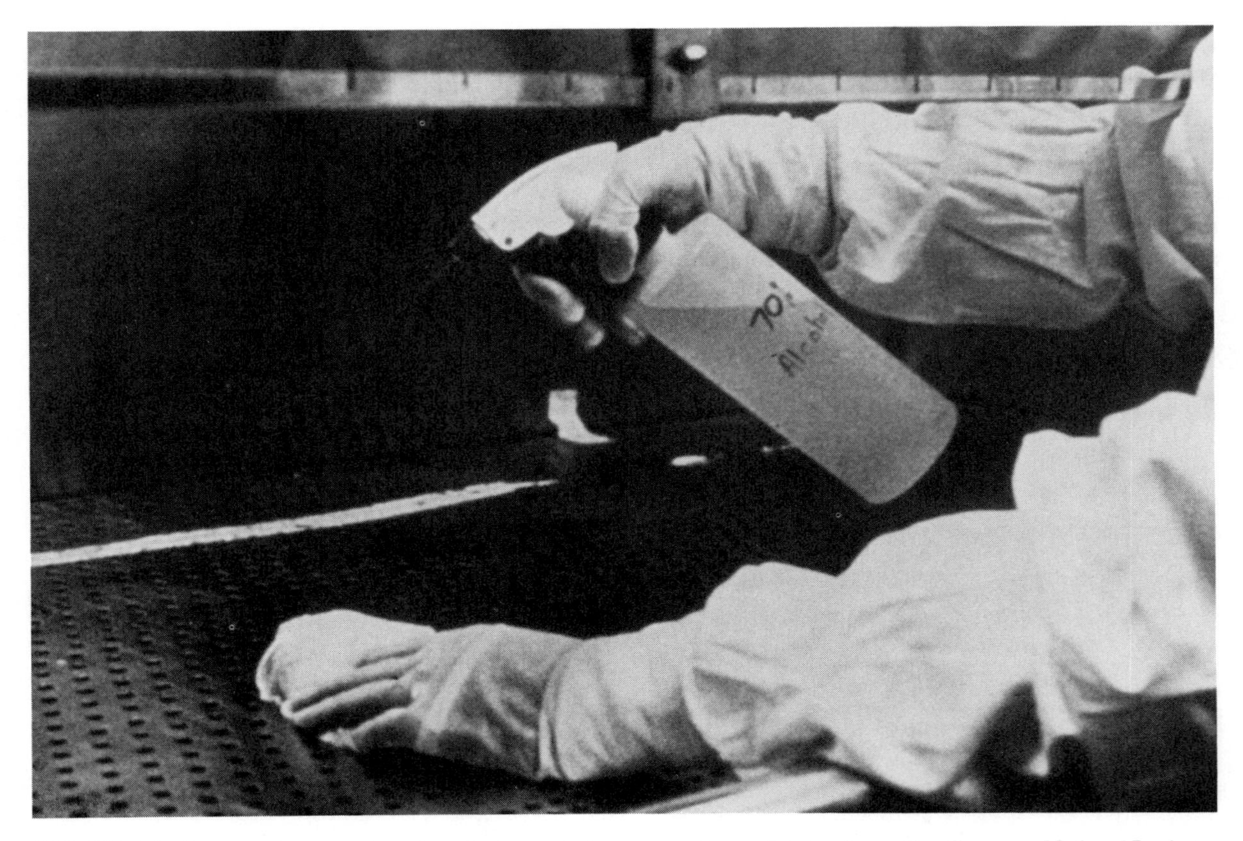

FIGURE 1.4 Decontaminate work surfaces before and after use, and immediately after spills. Courtesy, National Institutes of Health.

5. Labeling of Specimens

All clinical specimens should be regarded as infectious and handled accordingly, whether or not they bear special hazard warning labels. (See Chapter 3, Section H.)

6. Biosafety in Large-Scale Production

The principles and practices described for the control of biohazards in benchtop operations should be applied to the propagation of microorganisms in large-scale production processes. (See Chapter 3, Section K.)

7. Safe Disposal of Infectious Laboratory Waste

Just as with chemically hazardous waste, the generators of infectious laboratory waste have the responsibility to ensure that their waste is safely handled from "cradle to grave." Treatment and disposal of infectious waste by conventional methods such as solid waste incineration, sanitary landfills, and sewage treatment plants are adequate to protect the public's health. The risk of occupationally acquired illness, however, is present for workers who handle infectious waste. The following recommendations are presented to reduce this risk:

- **institutions should establish a waste management plan for the collection, segregation, containment, treatment, and disposal of infectious waste;**
- **workers involved in the handling of infectious waste should be informed of the potential hazard and should be proficient in the use of the necessary safeguards for their own protection;**
- **incineration, the method of choice for the decontamination of infectious waste, should be used whenever possible; and**

• **the principles of containment should be applied to all steps in the chain of handling infectious waste, from generation to disposal. (See Chapter 4.)**

8. Administrative Organization and Responsibility

The institution should have an overall safety, occupational health, and environmental protection program that includes biosafety in the laboratory and provides for compliance with federal, state, and local government regulations. The program should provide safety training for all laboratory, maintenance, and housekeeping personnel, for members of the medical staff, and for students who will come into contact with infectious materials in the course of their studies.

The responsibility for safety in a department or other administrative unit lies with its chairperson or supervisor. However, all individuals must accept responsibility for carrying out their work in a way that protects them and their fellow workers. Responsibility for safety must be clearly defined.

There should be a biosafety manual that sets forth the following: safety policy, goals, and organization; medical program; laboratory procedures; safety equipment; waste disposal methods; and emergency procedures. (See Chapters 4 and 5.)

9. Facilities

In order to provide an optimum environment for biohazardous work, facilities must be properly designed, constructed, validated, maintained, and used. Plans should be reviewed by the appropriate safety officer before construction begins, and again after the building has been completed but before work with infectious materials begins. Final construction (as-built) drawings should always be available for future reference. (See Chapter 5, Section B.)

10. Medical Program

Laboratories should have a medical program that is targeted to the potential risks and hazards of its operations. There should be a regular analysis of the collected data. Some employees may be found to have personal health conditions that place them at increased risk when performing certain laboratory functions. In such cases, a careful assessment should be made of the potential for exposure; if the exposure cannot be eliminated or reduced, consideration should be given to changing jobs or tasks. Employees should be informed completely about the potential risks and, to the extent possible, should be involved in the decision-making process. (See Chapter 5, Section D.)

11. Emergencies

Every laboratory organization has the responsibility to establish a specific emergency plan for its facilities and to be equipped to implement it. The plan should include the laboratory building as well as individual laboratories. For the building, the plan should include evacuation routes, shelter areas, medical treatment, and procedures for reporting accidents and emergencies. It should be reinforced by drills and simulated emergencies, and should include liaison with local emergency groups as well as community officials. For individual laboratories, the plan should cover such events as spills, fires, equipment failure, and accidents. (See Chapter 5, Section E.)

2

Descriptive Epidemiology of Occupational Infections of Laboratory Workers

A. INTRODUCTION

The precise incidence of occupational infections among laboratory workers is not known. During the past century, however, an extensive literature has documented that such infections have occurred with regularity and have occasionally resulted in death. The overall mortality rate of reported cases is 4 percent [100]. Published reports have dealt largely with single cases or outbreaks, retrospective studies, and passively collected anecdotal information. For the most part, historical accounts of laboratory-associated infections have listed only individual cases with no attempt at the difficult task of determining the size of the populations at risk. Furthermore, there is no central focus of responsibility or authority in the United States that maintains a comprehensive data base or conducts surveillance of occupational infections in laboratory workers who regularly or occasionally handle microorganisms or viruses.

B. THE EPIDEMIOLOGIC TRIAD

The components of the epidemiologic triad associated with laboratory-acquired infections are the host, the infectious agent, and the environment. Although most of the data relevant to these three factors have been collected retrospectively and are incomplete, an evaluation of the information that is available provides some insight into the complexity of the problem.

1. The Host

Comprehensive and current data are not available on the demography of laboratory workers (the host) who risk occupational infection with the agents they handle in their daily activities. Published surveys [28], however, indicate that the number of workers employed by public health and clinical laboratories is substantial, having been estimated at 250,000 in 1977. A more recent survey conducted in 1983 by the Occupational Safety and Health Administration (OSHA) [138] estimated that there are 370,000 employees in clinical laboratories, 45,000 employees in federal government laboratories, and 127,000 employees in academic laboratories. Additionally, the number of physicians' office laboratories in which one or more persons are employed may exceed 100,000 facilities. Of the combined total of more than 640,000 workers, as many as 500,000 may regularly or occasionally work with infectious agents or with blood, serum, urine, or other body fluids or tissues that may contain an infectious agent.

The surveys cited above do not include all segments of the general population of laboratory workers at potential risk of occupational exposure to infectious agents or their toxic or sensitizing metabolic products. Among the segments inadvertently excluded are a substantial but unknown number of persons whose duties may involve either regular or occasional handling of infectious materials: e.g., those persons working in animal and avian disease diagnostic or research laboratories, environmental laboratories, industrial and biologics production laboratories, forensic laboratories, and in laboratory animal production and care facilities. Consequently, the estimate of 500,000 workers who may be at risk of occupational infections probably represents a significant underestimation of the true number.

2. The Infectious Agent

Data are also not uniformly available about the

second epidemiologic component, the infectious agent. Details about the kinds of infectious agents encountered in the various categories of laboratories and the frequency with which such agents are handled are unknown. National morbidity data [31] indicate that hepatitis B virus (HBV) infections are widespread in the general population and that a pool of new cases and chronic carriers numbering several million may exist. For example, 1.0 to 1.5 percent of all admissions to large urban hospitals are positive for hepatitis B surface antigen (HBsAg) [45]. Although blood samples from these patients are potentially infectious, their identity is usually unknown throughout the hospitalization. These data suggest that HBV represents the specific infectious agent most likely to be transmitted in the clinical laboratory occupational setting.

3. The Environment

The third component of the epidemiologic triad, the laboratory environment, is also poorly defined. The physical features, including safety equipment and adequacy of the facility, vary widely. Although a number of federal, state, and private sector organizations (e.g., Health Care Financing Administration, OSHA, College of American Pathologists, Joint Commission on the Accreditation of Healthcare Organizations, American Association of Blood Banks, and various state agencies) do regulate, license, accredit, or inspect clinical laboratories, there is no consensus among these various organizations about the standards for laboratories operating under their respective jurisdictions. There is a current voluntary national code of laboratory practice [105] that describes features of a laboratory facility recommended for work with infectious agents, but there is no national authority that regulates laboratory operations conducted solely on an intrastate basis.

C. LABORATORY-ASSOCIATED INFECTIONS

1. Infectious Agents Presenting the Highest Risk

The series of survey summaries of laboratory-associated infections described by Pike in 1976 [100] indicates that, as a group, bacterial infections were the most frequently reported occupationally associated infections of laboratory workers during the first seven and one-half decades of the twentieth century. Viral and rickettsial infections were more frequently reported during the latter half of this time period. Of the 3921 cases reported, the five most frequently recorded infections in rank order were: brucellosis, Q fever, typhoid fever, viral hepatitis (all types), and tuberculosis. Most of these diseases were prevalent and important public health problems in our recent past, and their etiologic agents were handled commonly in clinical and diagnostic laboratories at that time.

While the compilations of laboratory-associated infections cited above provide a historical perspective of the hazards of occupational infection, these data are not necessarily indicative of present-day risk of infection. For example, brucellosis, the most frequently reported occupational infection, was formerly a major and widespread disease in human and animal populations. More than 6000 human cases were reported in the United States in 1947 [30]. As the incidence of brucellosis declined in domestic animal reservoirs, there was a corresponding decline in cases in the human population. By 1963, the annual number of cases reported had declined to 407 [43]. Of this number, only one case, a *Brucella suis* infection, was specifically identified as being laboratory associated.

Sharp decreases have been observed also in the annual number of reported cases of typhoid fever. In 1942, there were almost 6000 cases; in 1952, 2341 cases; and in 1984, 390 cases, of which approximately 70 percent were acquired during foreign travel [30]. Similarly, reported cases of tuberculosis decreased from 121,000 in 1950 to 22,500 in 1984 [30]. As the incidence of the "top five" diseases has decreased in the general population, there has been generally a corresponding decrease in the number of laboratory specimens received and examined that contain these agents. This decrease in numbers of specimens containing the infectious agents has certainly reduced the probability of occupational exposure. On the other hand, newly emerging diseases or newly recognized organisms may increase the risks of infection for laboratory personnel.

Tuberculosis is an interesting example of the latter situation. The recognition of the existence of species of mycobacteria other than *Mycobacterium tuberculosis*, *M. bovis*, and *M. avium* has required that specimens suspected of containing mycobacteria be subjected to increased laboratory manipulations, in order to recover as well as to identify the organisms. Such increased manipulations are not without inherent risks to laboratory personnel. The recent report of Miller et al. [84] summarizes the collective experience of several investigators and shows that tuberculosis has ranked among the top six causes of laboratory-associated infections for more than 25 years. These published reports probably represent only the "tip-of-the-iceberg" of laboratory-associated cases of tuberculosis. For example, the laboratory tuberculosis consultant at the Centers for Disease Control (CDC) has, at the time of this writing, been asked to assist in 13 investigations of laboratory-associated tuberculous infection, none of which has been published [74]. In these 13 investigations, 72 of 275 (26 percent) exposed individuals were found to have been infected with tubercle bacilli (i.e., recent tuberculin conversion); if untreated, 10 percent of these individuals would be expected to develop active tuberculosis. The fraction of exposed personnel infected in the different outbreaks ranged from 19 to 55 percent. Airborne dissemination was suspected in all instances.

It has recently been recognized that both acquired immunodeficiency syndrome (AIDS) [32,33,119,122] and intravenous drug abuse [36] may be contributing factors to the recent increase in morbidity from tuberculosis. Other factors that are suspected of playing a role are the influx of immigrants from Central America and the increasing numbers of homeless people. These factors can only cause an increase in the risk of mycobacterial infection for laboratory personnel. Despite excellent published surveys of the frequency of such infections, it appears obvious that many cases occur that are not reported in the medical literature.

Q fever differs from the other "top five" occupational infections in that this rickettsial disease has remained a relatively obscure public health problem of unknown incidence and sporadic distribution. However, Q fever is a proven and continuing hazard in those few facilities in which work with infected animals or human tissues is conducted, or in which the agent is propagated. This rickettsial agent is remarkably resistant to dessication and inactivation, and 10 or fewer organisms may produce infection via the respiratory route [142].

Hepatitis of varied etiology continues to be a community as well as an occupational health hazard among certain high-risk groups, including laboratory workers. Over the past 15 years, the incidence of HBV infections has shown a progressive annual increase, while in the same period, the number of reported cases of hepatitis A virus (HAV) infection has decreased [31]. In 1983, for the first time, the number of HBV cases exceeded those caused by HAV. While laboratory hazards of HAV infection are restricted primarily to persons working with experimentally or naturally infected chimpanzees, HBV poses a persistent and continuing hazard to all categories of laboratory workers handling clinical specimens of human origin. HBV has been demonstrated in a wide range of body fluids and tissues typical of those received and handled in clinical laboratories. The number of infectious virus particles may reach concentrations in excess of 100,000,000 per milliliter of blood. Since the early 1970s, when procedures for the serologic differentiation of HAV and HBV came into general use, HBV has become the leading cause of occupationally acquired infection among laboratory and health care workers [81]. Studies by Jacobsen and co-workers in Utah demonstrated a prevalence of clinical HBV infection in clinical laboratory personnel that was 14 times greater than that in the general population, i.e., 0.84 cases/100,000 versus 0.06 cases/100,000, respectively [70]. Elevated HBV infection rates or incidences of serologic markers were demonstrated in public health laboratory workers in the United Kingdom [64], in clinical chemistry workers in Denmark [114], and in small rural hospitals [45], as well as in large urban hospitals in the United States [66]. Osterholm and Andrews [97] have demonstrated annual infection rates for HBV and non-A/non-B hepatitis that exceeded 10,500 cases/100,000 in the staffs of hospital dialysis units, while employees in nondialysis units of these hospitals exhibited a rate of approximately 500 cases/100,000. The ratio of HBV to non-A/non-B hepatitis infections was greater than 5 to 1.

Lauer found HBV antigen on one-third of the surfaces of work areas, equipment, and laboratory implements sampled in a large, modern medical center laboratory [76]. Collins found visible blood on the labels of 17 percent of tubes of blood specimens, and feces on the external surfaces of 6 percent of the containers of the stool samples received at a public health laboratory in England. Four to five percent of laboratory services request forms received by another public health laboratory in England were visibly blood stained [42]. Bond et al. demonstrated that HBV remains viable and infectious after being dried in serum and held for seven days at ambient laboratory environmental conditions of 25°C and 42 percent relative humidity [20].

The primary routes of occupational infection with HBV, in rank order, are as follows: accidental parenteral self-inoculation with infectious fluids (needle sticks) [45,81]; exposure of the mucous membranes of the eyes, nose, or mouth to infectious materials; and, possibly, contamination of the skin with infectious materials.

In a joint advisory notice, the U.S. Departments of Labor, and Health and Human Services, have informed employers about the serious occupational infection problems of HBV, human immunodeficiency virus (HIV), and other blood-borne diseases [139]. In this advisory, federal health officials estimated that as many as 18,000 health-care workers may be infected in a single year with HBV. Of these cases, as many as 12,000 may be occupationally associated. It was further estimated that nearly 10 percent of the cases will become long-term carriers of the virus, and that more than 200 health care workers may die as the result of the HBV infection or associated complications.

The evidence is overwhelming that, of all indigenous pathogens, HBV has the greatest potential for transmission within the occupational setting of the clinical laboratory. This conclusion is based upon the comparatively high frequency of asymptomatic carriers, the high titers of virus in blood and other body fluids, the stability of the virus on work surfaces and other items in the laboratory, the low infectious dose, the multiple routes of infection, and the demonstrated occupational incidence of infection.

An essential consideration in the occupational risk assessment of HBV and other infectious agents for which only Biosafety Level 2 (see Appendix A) practices are recommended is the lack of evidence suggesting that occupational transmission occurs by means of true infectious aerosols, i.e., inhalation of respirable particulates typically less than 5 microns in diameter.

While not among the most prevalent occupational infections recorded by Pike (i.e., the "top five") [100], shigellosis is historically and currently a continuing occupational risk. The low oral infectious dose (on the order of 100 viable organisms) facilitates transmission in the occupational setting, as well as in the general population [142]. In a retrospective study of more than 20,000 British medical laboratory workers, shigellosis was the third most frequently recorded occupational infection, following viral hepatitis and tuberculosis [64].

2. Infectious Agents Presenting the Lowest Risk

In contrast to the proven occupational infection hazard of HBV and the other "top five" agents, a number of infectious agents handled in laboratories have exhibited a consistent history of remarkably low incidence or absence of reported occupational infections. Examples of such agents include rabies virus, Creutzfeldt-Jakob agent (CJA), *Vibrio cholerae, Clostridium tetani,* and HIV. While the consequences of infection with any of these five agents are serious, the cumulative history of laboratory experience attests to the low risk of transmission in the laboratory setting.

In the almost 100-year history of work with rabies virus in diagnostic and research laboratories, often in the most primitive of facilities and without preexposure immunization of personnel, only two documented cases of laboratory-associated infections have been recorded. Both cases occurred under conditions involving the manipulation of relatively large quantities of high-titer virus suspensions: one in a production facility [27] and the other in a research laboratory [29]. Exposure of personnel to aerosols of high-titer virus suspensions was the most plausible explanation for each of these two unusual cases. Neither the quantity nor the concentration of the virus in the materials handled, nor the procedures performed, was typical of the conditions in a diagnostic or clinical laboratory.

Creutzfeldt-Jakob agent, a slow virus causing transmissible viral dementia, is an infectious agent that can be passed serially from human to human, and from human to susceptible nonhuman primates or rodents. Extensive experience with CJA and the clinical disease (CJD) up until the present time has indicated that "none of the people in closest contact with patients with CJD (wives, friends, employee contacts, members of the medical or nursing professions, or paramedical personnel) appears to have a higher risk of contracting CJD than does the general population. Not a single case of CJD has yet been reported to have occurred in workers most exposed to infectious tissues from patients with CJD (neuropathologists, research scientists, and laboratory personnel). Thus, despite proven person-to-person transmissibility of the disease by invasive procedures, the risk of acquiring CJD by any means other than tissue penetration by contaminated materials must be very small indeed" [23]. It should be noted that two accounts of the occurrence of CJD in laboratory workers were recently published, although the causal relationship between the disease and occupational exposure was not established in either case [85,113].

While cholera is periodically epidemic in tropical and subtropical countries, only 12 laboratory-associated infections have been reported during this century [100]. The very high oral dose required for infection, of the order of 100,000,000 viable organisms [142], is undoubtedly a major reason for the small number of laboratory-associated cases.

Although Pike recorded five laboratory exposures to toxin of *Clostridium tetani* produced in vitro [100], there have been no recorded cases in laboratory workers of occupational infections or intoxications with *C. tetani*, *C. botulinum*, or their respective toxins.

Few infectious agents have generated more concern and anxiety over potential occupational exposure and hazards of infection than has HIV. Active prospective surveillance, however, has shown that fewer than 1 percent of overt exposures (including needle sticks) of people attending patients with AIDS or with other manifestations of HIV infection, have resulted in seroconversion of the exposed individuals [35]. The majority of those health care workers with reported occupationally acquired HIV infection have a history of needle stick exposure to blood of infected patients in the clinical setting. As of February 1988, there have been three reported seroconversions in laboratory workers. One of these three cases occurred in a medical technologist who spilled blood from an infected patient on her ungloved hands and forearms while manipulating an apheresis machine [35]. The other two recorded cases occurred in employees of large-scale virus production facilities propagating HIV for research or reagent use [36,143]. One of these two workers had an overt parenteral exposure to a concentrated virus preparation. The other worker had no recognized accidental occupational exposure or any risk behavior linked to HIV infection.

Despite the low incidence of transmission in the laboratory, the potentially life-threatening consequences of HIV infection mandates that all laboratory workers who handle blood, body fluids, tissues, or cultures utilize those laboratory practices and personal protective measures identified as "Universal Precautions" by the CDC, which are recommended for the prevention of transmission of HIV and other blood-borne diseases [34].

Common to each of these infectious agents with a demonstrated low risk of occupational infection is the fact that primary occupational infection is associated with one of the following exposures: accidental parenteral inoculation (e.g., needle stick); contamination of the mucous membranes of the eyes, nose, or mouth with infectious droplets (particulates typically greater than 5 microns in diameter); ingestion; or penetration of the intact or broken skin by the agent. There is no documented risk of transmission by means of an infectious aerosol (particulates typically less than 5 microns in diameter) generated during the manipulation of clinical materials or of diagnostic quantities of the agent.

3. Other Infectious Agents

Except for some exotic microbial agents, the occupational risk of infection with virtually any biological agent falls between the extremes observed with the "top five" and the "low-risk" groups of infectious agents described above. The recommended facilities, equipment, and microbiological practices necessary for the handling of infectious agents are detailed in Chapter 3 and in Appendix A, which is a reprinting of the Centers for Disease Control (CDC)/ National Institutes of Health (NIH) publication, *Biosafety in Microbiological and Biomedical Laboratories* [105]. These guidelines should be followed when contemplating work with any potentially infectious agent. Recommendations for handling HIV, an infectious agent that was identified after the CDC/ NIH publication, are reproduced in Appendixes B and C [34,36,38].

3

Safe Handling of Infectious Agents

A. GUIDELINES FOR HANDLING PATHOGENIC MICROORGANISMS

In 1984, the Centers for Disease Control (CDC) and the National Institutes of Health (NIH) jointly published a set of guidelines for the safe handling of pathogenic microorganisms [105]. These guidelines, developed over a period of several years in consultation with experts in the field, remain the best judgments available; they are reproduced here in their entirety, as Appendix A. The reader should consult these guidelines in deciding on the appropriate level of precaution to use in the handling of a particular organism.

Guidelines for handling agents identified after the CDC/NIH publication are published as Agent Summary Statements in *Morbidity and Mortality Weekly Report* (MMWR), issued by the CDC. The Agent Summary Statement for human immunodeficiency virus (HIV) [36] is reprinted here as Appendix B, and additional MMWR articles on HIV ("Recommendations for Prevention of HIV Transmission in Health-Care Settings" [34,38]) are reprinted here as Appendix C.

Throughout this and the following chapters, frequent reference is made to Biosafety Levels 1 through 4. These levels are described in the CDC/NIH publication (Appendix A). Table A.1 of this appendix summarizes the practices, techniques, and safety equipment prescribed for each level.

B. ORGANISMS POSING SPECIAL RISKS

The risk of acquiring an infection in the laboratory is influenced by many variables. Among these factors are the health and immune status of the laboratory worker, the suitability of the laboratory for work with highly pathogenic agents, the characteristics and the concentrations of the microbe being handled, and the specific manipulations involved in its handling.

Studies of infections acquired by personnel working in microbiological laboratories have been carried out by several investigators over the past half-century [42,84,101,105,120,121] and have identified a number of potential human pathogens that are clearly more frequent causes of laboratory-acquired illnesses than are others (see Chapter 2, above). Organisms falling in this category are to be found among viruses, bacteria, rickettsiae, and fungi. Awareness of those species with a high potential for invading normal humans should lead to the use of appropriate precautions to minimize the risk of infection.

Among the agents that have been identified in recent years as posing the greatest risk of infection to laboratory and ancillary personnel of diagnostic laboratories are the virus of hepatitis B, *Mycobacterium tuberculosis,* and *Shigella* spp. [60,70,121]. A partial list of other agents known to pose greater than average risk to laboratory workers includes *Brucella* spp., *Salmonella* spp., leptospires, *Coxiella burnetii, Rickettsia* spp., and *Coccidioides immitis.* The recently identified virus of AIDS (HIV), on the other hand, poses a low risk of occupational infection to laboratory workers, except to those working with concentrated virus suspensions [37,143]. The supplement to the CDC/NIH guidelines recommends, therefore, that HIVs be handled according to the standards and special practices of Biosafety Level 2 or 3, depending on the concentration or quantity of virus or the type of laboratory procedure used (see Appendix B).

No agent that is a component of the normal or abnormal microbial flora of man should be regarded as lacking totally in pathogenic potential, and all microorganisms should be handled with appropriate techniques. With the increase in research in virology in the past half-century, laboratory infections with viruses have increased relative to those caused by bacteria and mycoplasmas.

An important defense against infection with some viral agents is immunity induced by vaccination. Whenever a vaccine is available (see Table 5.2), its use should considered for those at risk of exposure prior to their handling of the virus in question. Under certain circumstances, when work with highly virulent agents is contemplated, it may be necessary to consider the administration of an experimental vaccine. Because of the potential risk of injury to the fetus from apparent or inapparent viral infection, special precautions, including temporary reassignment, may be considered for female personnel who are pregnant or are contemplating pregnancy. (See Chapter 5, Section D.)

All personnel working with infectious agents should have documented evidence of immunization with the vaccines required by most jurisdictions for admission to elementary school, e.g., diphtheria, tetanus, pertussis, poliomyelitis, measles, mumps, and rubella. In addition, vaccines for preventing infections with other agents to which they may be exposed, if available, should be offered, and in certain circumstances consideration should be given to making such immunization mandatory.

Acceptance of immunization against, or demonstration of proven immunity to, hepatitis B virus should be a precondition for the employment of all workers who will be handling human blood or body fluids. If the medical program of the hiring organization includes a serum bank, a sample should be obtained at the time of employment and stored in the frozen state, to provide a baseline for subsequent immunologic assays as required (see Chapter 5, Section D).

C. HAZARDS FROM VERTEBRATE ANIMALS AND INSECTS IN THE LABORATORY

Personnel who work with experimental vertebrate animals in the laboratory, or who receive and handle specimens from vertebrate animals, should be cognizant of the potential for exposure to zoonotic pathogens and to allergenic animal danders, urine, and saliva.

A list of zoonotic pathogens and potential animal sources of infection for humans is included in Appendix D; the information for this table was derived from references 6, 17, 21, 26, 53, and 69. While it is recognized that many of the agents listed are not significant hazards under ordinary laboratory circumstances, laboratory staffs should recognize the dangers of zoonotic pathogens and should realize, for example, that protozoan cysts and larval stages of certain helminths in fecal material can be infectious [26]. Application of the seven basic rules of biosafety cited in Section F of this chapter will greatly reduce the risks of infection while handling vertebrate animals or specimens obtained from them (see also Section G of this chapter).

Strong consideration should be given to immunizing employees with appropriate vaccines against zoonotic agents, if available (see Table 5.2).

Numerous agricultural, veterinary, and human disease research laboratories are involved in the production and maintenance of insects. Insects are also produced for regulatory and control activities (e.g., screwworm control, which involves the release of insects into the environment). The human health hazards of insect production have been recognized recently. In addition to the hazards associated with insect bites, allergic reactions and respiratory diseases may result from contact with, or aerosol exposure to, various insect developmental stages, insect waste products (e.g., body hairs and feces), ingredients used in insect diets, or mold spores and bacteria that contaminate larval diets. Repeated exposure over a period of months or years may produce respiratory ailments or other manifestations of allergic reactions in susceptible individuals.

During the preplacement medical evaluation at the time of hiring or job assignment, a history of allergies to vertebrate animals or insects that the prospective employee is likely to encounter should be elicited. After hiring, the periodic monitoring medical examinations should include an evaluation for the development of allergies (see Chapter 5, Section D). The prevalence of allergies among personnel who work with or are exposed to vertebrate laboratory animals has been estimated to be 11 to 30 percent [14]. Some individuals may become very sensitive to low concentrations of allergens [150,151]. More than 300 cases of allergic reactions that probably resulted from the inhalation of insect-derived materials have been reported [11]. More than 40 species (among eight orders) of insects were associated with work-related allergic symptoms among U.S. Department of Agriculture employees working with insects [10]. Insect allergy questionnaires and surveys indicate that respiratory symptoms (e.g., sneezing, coughing, and chest tightness) and eye and skin irritation or skin rash are the major symptoms in those with complaints of insect allergy [21,146]. Inhalation of airborne material was reported as the mechanism most frequently responsible for allergic symptoms in persons working in insect-rearing facilities [21,146].

Most insect-related health problems develop after repeated exposure, and severity often increases with continued exposure. Sensitivity and susceptibility vary greatly among individuals. The allergic symptoms of conjunctivitis, rhinitis, sinusitis, asthma, or pruritus and dermatitis can develop in from less than one year to many years after initial exposure. Whether or not people with allergies are more likely to develop additional allergies to animal products is controversial. Precluding allergic individuals from employment does not eliminate the problem, since nonallergic individuals also can become sensitized.

Reducing contamination levels and reducing exposure are the best preventive measures. This may be accomplished by engineering controls such as filtration and directional control of airflow, or by the use of filter-top cages and directional airflow racks to prevent the allergens from reaching the worker. The selection, design, and utilization of such equipment are the most important steps in controlling respiratory hazards. Respirators should be used only for temporary or intermittent work, such as during maintenance work on the ventilation equipment, and should not be relied upon as a permanent solution [151]. It may be appropriate for vertebrate animal caretakers, insect production workers, laboratory personnel, and others who work with animals or who enter the animal holding areas to wear gloves, eye protection, and a mask covering the nose and mouth. It is good practice to change from street clothing to laboratory garb. All persons who enter the animal holding area should adhere to the protocols and the regulations that apply to activities in the vivarium.

D. PRIMARY AND CONTINUOUS CELL CULTURES

Cell cultures, in general, present few biohazards in the laboratory, as evidenced by their extremely wide usage and the rare cases of transmitted infections to laboratory personnel. Primary cell cultures initiated with tissues from infected humans or animals are recognized hazards. Thus macaques, and possibly other Old World monkeys, may have latent *Herpesvirus simiae* (B-virus) infections and present a hazard to personnel handling these animals and their tissues. At least 24 documented cases of infections of laboratory workers handling primary cell culture tissues (e.g., primary rhesus monkey kidney cells) have occurred in the past 30 years [46]. A particularly noteworthy instance of the laboratory infection of a number of workers by an adventitious agent from monkeys occurred in 1967 in Marburg and Frankfurt, Germany, and in Yugoslavia. Laboratory workers handling tissues and cell cultures from African green monkeys developed an acute febrile illness. Seven deaths occurred among 31 documented cases due to a previously unknown virus, subsequently named Marburg virus. It has not occurred in laboratory workers since those incidents [115]. Tissues from mice infected with lymphocytic choriomeningitis (LCM) virus or from chickens carrying Newcastle Disease virus (NDV) also present potential hazards, but such laboratory infections have not been reported. Clearly, primary cell cultures prepared from humans infected with hazardous agents (e.g., HIV) present danger of infection, and such tissues must be handled with the precautions required of the known or suspected infectious agent (see Appendix A).

Continuous cell cultures present no real documented risk in the laboratory unless they are carelessly contaminated with an infectious agent. All continuous cell lines should be regularly monitored for contamination with infectious agents, and it should be emphasized that all nutrient media or other reagents that may contain ingredients of biologic origin must be treated as though they contain potentially infectious agents.

E. HANDLING OF NECROPSY AND SURGICAL SPECIMENS

1. Introduction

Necropsy and surgical pathology expose health care workers to various infectious agents that may be in human tissues or associated body fluids. Proper handling can minimize the risk of infection. Because the consequences of infection are grave, agents of principal concern are hepatitis B virus (HBV), human immunodeficiency virus (HIV), Creutzfeldt-Jakob agent (CJA), and *M. tuberculosis*, although a number of other infectious agents, including viruses, rickettsiae, bacteria, fungi, and parasites, pose potential risks. The principal means of acquiring infections when performing anatomic examinations are through breaks in the skin caused by needle punctures, cuts, or severe dermatitis, by contamination of mucous membranes, and by inhalation. The risk of infection is decreased by preventing breaks in the body surfaces, preventing the formation of droplets that might contaminate surface breaks or mucous membranes, inserting barriers such as rubber gloves, goggles, and masks between the infectious hazard and the potential site of entry, and preventing the generation of aerosols.

Fresh tissue may be infected with agents such as HBV or HIV even if there is no history of such infection. Those who perform autopsies or handle fresh tissue or blood on a regular basis should have immunity to HBV.

To date, in the United States there are few recommendations for biosafety in necropsy and surgical pathology [9,79,90], although such have been proposed in Great Britain [47]. Recommendations have been developed for Creutzfeldt-Jakob disease (CJD) [109], and a number of publications have helped to define the risks associated with this agent [9,24,55,56]. The National Committee for Clinical Laboratory Standards (NCCLS) has published proposed guidelines, entitled *Protection of Laboratory Workers from Infectious Disease Transmitted by Blood and Tissue,* that include necropsy and tissue handling recommendations [90]. Independently, a committee of the College of American Pathologists is developing recommendations for necropsy and surgical pathology.

In a given institution, there should be a clear definition of the responsibility for biosafety in the handling of a body from the time of death until it is transferred to the mortician or incinerated, and for surgically removed tissue. For an autopsy, the prosector (generally a pathologist) is responsible for biosafety. It is beyond the scope of this publication to discuss autopsy biosafety in detail, but some major points are considered below.

2. Necropsy

a. Routine Necropsies

Because of the high incidence of asymptomatic carriers of HBV and HIV in hospital or forensic autopsies, all cases should be considered potentially infectious and the necropsy performed carefully. Care should be taken to minimize chances of needle sticks, cuts, or abrasions. Risk of contamination of mucous membranes should be decreased by wearing safety goggles and a surgical mask, or a face shield.

b. Necropsies on Bodies Known to Be Infected

Bodies for necropsy should be appropriately labeled if they are known to be infected with such agents as HBV, CJA, HIV, or *M. tuberculosis*. The medical record should also indicate the diagnosis. Before beginning a dissection, it may be helpful to discuss the case with the clinician to clarify the extent of examination required. Autopsy assistants should be informed of the nature of the clinical diagnosis so that special disinfectants, such as sodium hypochlorite solution (household bleach diluted 1:100 in tap water), can be prepared prior to beginning the dissection.

In addition to the prosector and autopsy assistant, it is helpful to have a "circulating" assistant who remains "uncontaminated," thus preventing contamination of telephones, cameras, drawer pulls, cultures, papers, and other items by those doing the dissection, and confining the contamination to the necropsy table area.

Protective clothing should include the following:

- **a scrub suit covered with a long-sleeved gown or a long-sleeved coverall suit plus an impervious apron;**
- **impervious shoe covers;**
- **head covering;**
- **goggles or eyeglasses to prevent conjunctival contamination;**

- **face mask to decrease risk of droplet contamination of mucous membranes, or inhalation of aerosols; and**
- **double gloves (preferably including one pair of heavy-duty gloves).**

In performing the autopsy, it may be helpful to cover rib ends with towels to decrease risk of cuts. Dissection in the body should be limited to one prosector at a time. Use of scissors when possible will decrease the risks of cuts. Production of droplets and aerosols should be minimized. Use of a Stryker saw to open the skull or to cut bone is controversial because of the potential for generation of droplets and aerosols. Some authorities advocate using a hand saw, whereas others recommend using the Stryker saw with a HEPA-filtered vacuum attachment or covering the equipment with a wet towel. The saw and aerosol control apparatus should be adequately disinfected after use. In cases of CJD, there should be special care not to cut the brain. A new technique for the removal of the brain from cases of AIDS at autopsy has been developed in which the sawing is done inside a plastic bag [78,79]. The British Committee on Dangerous Pathogens has suggested performing limited postmortem examinations with discrete tissue sampling for most AIDS cases [1].

Any spills of blood or body fluid should be cleaned immediately with a solution of household bleach diluted 1:100 in tap water.

Specimens for culture or other clinical laboratory examinations should be handled in the same fashion as in patient care areas, with care being taken not to contaminate the outside of the container.

Disposable syringes and needles and knives should be placed in a leak- and puncture-resistant container for subsequent disposal.

If persons are cut or punctured while dissecting or handling tissues or body fluids, the wound should be encouraged to bleed, flushed with abundant water, and treated with an antiseptic such as povidone-iodine. The accident should be reported to the appropriate persons such as the safety officer, employee health director, or laboratory supervisor, depending on the institutional requirements.

At completion of the autopsy, the body should be packed with absorbent material to prevent seepage of liquids and should be washed with a 1:100 dilution of household bleach or other appropriate disinfecting agent. Tags on the body should note the infectious hazard. The body should then be placed in a plastic bag, which is also labeled with the appropriate hazard warning (e.g., "Blood and Body Fluid Precautions"). In addition to labels on the body, the mortician should be notified specifically of the infection hazard. As discussed below in Section F of this chapter, however, the use of special hazard warning labels should not lead to the misconception that other bodies are not potentially infectious.

When finished, prosectors and autopsy assistants should remove protective clothing in the autopsy room and place it in appropriate containers for incineration or transport to the isolation laundry, and should then shower. Soiled disposable items should be placed in biohazard bags for incineration. Soiled linens should be double-bagged in durable, labeled isolation bags and handled in the same manner as hospital isolation linen.

Tissues that are to be saved should be placed in formalin (1 part tissue to 10 parts formalin) and should be cut thin enough (<2 cm thickness) to ensure penetration. Fixation in 10 percent formalin will inactivate most infectious agents; mycobacteria and CJA are exceptions (see below).

Instruments should be autoclaved or soaked in a 1:100 dilution of household bleach, or other appropriate disinfectant, for 30 minutes to 1 hour. Only stainless steel can be placed in hypochlorite solution. The table and the floor around the table should be cleaned with a 1:100 dilution of household bleach, or with a germicide approved (by FDA) for use as a "hospital disinfectant" that is also tuberculocidal. If a mop is used, it should be autoclaved.

Creutzfeldt-Jakob agent is particularly resistant to killing, requiring autoclaving at 121°C for at least 30 minutes; it can survive in 10 percent formalin for many months [9,56]. Paraffin blocks may therefore contain infectious CJA. CJA is usually inactivated by household bleach at 0.5 to 5 percent concentrations, with the higher concentration being more effective but also more corrosive [24]. The agent is most susceptible to 1N NaOH. Contaminated material should be autoclaved as above, inactivated with one of the chemicals cited above, or incinerated. It has recently been noted that formalin-fixed brain tissue can be autoclaved to inactivate CJA and then processed for histologic sections [80].

HIV and HBV are readily inactivated by a variety of agents, including formalin, hypochlorite, and iodine-based disinfectants. Special care should be exercised when performing autopsies on patients who died of infections with these agents.

Mycobacterium tuberculosis is moderately resistant to 10 percent formalin, requiring prolonged exposure for complete killing [110]; formalin-fixed tissue from recent cases may therefore be infective. The usual route of infection is the inhalation of aerosols generated during necropsy, or the trimming of tissue for histologic processing. Occasionally, the organism is introduced into a cut ("prosector's wart").

3. Surgical Pathology

The hazards of surgical pathology are similar to those of autopsy. Many tissues have been fixed in formalin when received and are thus not infectious, with the exceptions noted above. Such tissues are best disposed of by incineration, more for aesthetic reasons than those related to biohazard.

Cryostats used for frozen sections present a particular problem [123]. The operator should wear gloves, gown, and mask when cutting the section, whether or not the patient is known to have a disease transmitted by blood or tissue. In addition, the cryostat should be disinfected periodically (at least weekly). If it is known that the patient has an infection that represents a hazard such as AIDS or tuberculosis, frozen sections should be prepared only when absolutely necessary. The cryostat should be disinfected with an appropriate disinfectant as soon as possible after the sections have been cut, to remove contaminated tissue fragments and to decontaminate surfaces.

All human anatomical waste and cadavers should be disposed of by burial or incineration. The incinerator must be appropriately designed for handling anatomical laboratory waste. Cadavers containing radioactive isotopes or antineoplastic drugs require special handling during autopsy and for disposal (see Chapter 4).

F. GOOD LABORATORY PRACTICES

1. Introduction

A number of reports and studies [8,15,40,67, 68,77,84,96,101] attest to the potential for occupationally acquired infection by laboratory personnel working directly with microbial agents. The significant element to be derived from these reports is that the exact source or cause of the infection could be documented in fewer than 20 percent of the cases. This finding provides strong evidence that exposures and consequent infection occur not as the result of overt accidents but during the performance of routine procedures.

2. Routes of Exposure

The nature of infective contaminants dispersed during the performance of any laboratory procedure is a direct function of the amount of energy applied during the procedure. Low-energy procedures (e.g., removal of screw caps and pouring of liquid medium) principally yield droplets that are dispersed onto body and work surfaces. Exposure of personnel in these instances occurs usually through breaks in the skin surface caused by cuts, scratches, and other cutaneous lesions, or by ingestion of infectious material transferred to the mouth by hands or objects. On the other hand, procedures involving application of large amounts of energy, such as homogenization and centrifugation, have the potential for generating respirable aerosols. It should be recognized that a large number of procedures may result in the generation of a mixture of droplets and aerosols with the result that exposure by more than one route is possible.

While it has been typical to focus on respirable aerosols as the primary source of infection for laboratory personnel, it is essential that other routes of exposure be considered: contact, oral, ocular, and inoculation.

a. Contact Route

The control of potential exposure by the contact route requires that procedures be conducted in a manner that avoids contamination of body or work surfaces. This is accomplished through the use of gloves and other personal protective clothing, protection of work surfaces with appropriate absorbent disposable covering, use of care in the performance of procedures, and cleaning and disinfecting work surfaces. Procedures that can result in the generation of droplets include decanting of liquids, pipetting, removal of screw caps, vortex mixing of unsealed containers, streaking inocula on agar surfaces, and inoculation of animals.

It should be recognized also that dispersal of contaminants to other surfaces can occur by their transfer on the gloves of the laboratory worker, by the placement of contaminated equipment or laboratory ware, and by the improper packaging of contaminated waste.

b. Oral Route

A number of procedures carried out in the laboratory and animal facility offer the potential for either direct or indirect exposure by the oral route. The procedure that offers the greatest potential for exposure by ingestion is mouth pipetting. Clearly, such exposures are completely avoidable through the use of mechanical pipetting devices. Indirect oral exposures can be avoided through the use of the personal hygienic practice of regular hand washing, and by not placing any objects, including fingers, into the mouth. The wearing of a surgical mask or face shield will serve to protect the worker against the splashing of infectious material into the mouth.

c. Ocular Route

The wearing of a face shield, safety glasses, or goggles will protect the worker against splashing infectious material into the eyes.

d. Inoculation Route

The single procedure that presents the greatest risk of exposure through inoculation is the use of a needle and syringe. These are used principally for the transfer of materials from diaphragm-stoppered containers and for the inoculation of animals. Their use in the transfer of materials from diaphragm-stoppered containers can, in addition, result in the dispersal of infectious material onto surfaces and into the air. Depending upon the route of inoculation of animals, the use of a needle and syringe may also result in the contamination of their body surfaces. Because of the imminent hazard of self-inoculation, the use of the needle and syringe should be limited to those procedures where there is no alternative, and then the procedure should be conducted with the greatest of care. Inoculation can also result from animal bites and scratches.

e. Respiratory Route

Several procedures have the potential for generating respirable aerosols. Included are sonication, homogenization, centrifugation, vigorous discharge of fluids from pipettes, heating inoculating loops, opening lyophilized preparations, and changing of the litter in animal cages (see Chapter 3, Section I).

3. Prevention of Exposure

The time-honored approach for the safe handling of infectious agents involves the use of a combination of strategies. This is accomplished by

- **controlling the hazardous material at the source to prevent release into the workplace,**
- **minimizing accidental release of the material, and**
- **protecting the worker against contact with the material.**

However, the safe conduct of work with infectious material is primarily dependent upon the application of good laboratory practices by the laboratory worker (see below).

4. The Seven Basic Rules of Biosafety

The most common means of exposure can be essentially eliminated as occupational hazards by following the seven basic rules of biosafety:

- **Do not mouth pipette.**
- **Manipulate infectious fluids carefully to avoid spills and the production of aerosols and droplets.**
- **Restrict the use of needles and syringes to those procedures for which there are no alternatives; use needles, syringes, and other "sharps" carefully to avoid self-inoculation; and dispose of "sharps" in leak- and puncture-resistant containers.**
- **Use protective laboratory coats and gloves.**
- **Wash hands following all laboratory activities, following the removal of gloves, and immediately following contact with infectious materials.**
- **Decontaminate work surfaces before and after use, and immediately after spills.**
- **Do not eat, drink, store food, or smoke in the laboratory.**

These simple and effective work practices can be implemented readily by laboratory management at minimal cost and with no loss of employee efficiency or productivity. Even in the absence of more sophisticated means for providing safety in the laboratory, these practices can achieve a major reduction in the risk of accidental infection.

Laboratory activities that pose the risk of infection via airborne aerosols or droplets demand the use of special safeguards. For many "airborne pathogens," the human infectious dose may be as low as one viable microorganism, as demonstrated for tuberculosis [107,108]. It is recommended that biological safety cabinets or other primary containment devices be used for all manipulations of materials, including clinical specimens, known to contain or suspected of containing microorganisms capable of infecting by the respiratory route. In laboratories where such materials are handled, the ventilation system should provide directional airflow from "clean" to "contaminated" areas, and the air should not be recirculated.

The recommended procedures listed above, targeted at minimizing overt occupational exposures, constitute the basic essentials of good laboratory practice. Furthermore, these procedures are also effective in reducing or eliminating overt exposure to the variety of indigenous bacterial, viral, fungal, and parasitic agents present in the community and commonly found in clinical material submitted to the laboratory for examination. The ultimate responsibility for assessing the risk of occupational infections and for implementing appropriate practices, as well as for providing adequate facilities and containment equipment, rests with the laboratory director.

5. Summary

Virtually all laboratory procedures have the potential to disperse infectious material into the workplace. Laboratory workers should be aware of these potential hazards and exercise a high degree of care during all manipulations of infectious materials. As evidenced by the data accumulated in the review of laboratory-acquired infections by Pike [101], exposure of laboratory workers is not often associated with overt accidents. More than 80 percent of laboratory-associated infections could not be ascribed to any specific event. It is critical, therefore, that laboratory workers recognize that good microbiological practices are required to prevent exposure to infectious agents. These practices are described in more detail in subsequent sections of this chapter.

G. TRANSPORTATION AND SHIPMENT OF SPECIMENS

1. Introduction

Although it is obvious that biological specimens should be properly packaged, labeled, shipped, and received, concerned national and international organizations have found it necessary to develop recommendations and guidelines because of the fear of accidents and spills involving such materials [71,86,147,148]. Federal regulations govern the packaging and shipping of hazardous materials. The importation and subsequent transfer between laboratories of etiologic agents and vectors of plant, animal, and human diseases (including zoonotic agents) are controlled through permit systems.

2. Packaging. Shipping, and Handling of Biological Specimens

The shipment of diagnostic specimens, biological products, and etiologic agents concerns everyone involved in the process. Infectious materials that are properly packaged and handled may pose considerably lower risks of accidental exposure for nonlaboratory personnel who come in contact with the shipment in transit. Proper packaging also may ensure considerate and prompt handling of valuable specimens.

The shipping of unmarked and unidentified etiologic agents is prohibited. Requirements for the proper method of containment in the packaging and the use of the hazardous warning label are stipulated in the U.S. Public Health Service Interstate Shipment of Etiologic Agents Regulation [129]. Comparable requirements of the International Civil Aviation Organization (ICAO) apply to the international shipment of diagnostic specimens and infectious agents.

The containment packaging and hazard warning labeling specified in the U.S. Public Health Service Regulation [129] for the shipment of etiologic agents is illustrated below in Figure 3.1. The package should consist of

- **a securely closed, watertight primary container (test tube, vial, or ampoule);**
- **a durable, watertight secondary container; and**

ETIOLOGIC AGENTS

BIOMEDICAL MATERIAL

IN CASE OF DAMAGE OR LEAKAGE

NOTIFY DIRECTOR CDC ATLANTA, GEORGIA

404/633–5313

STANDARD FORM 420 JUNE 1973
PRESCRIBED BY DEPT HEW (4.2 CFR)
420–101

FIGURE 3.1 Containment packaging and hazard warning labeling specified by the U.S. Public Health Service. Reprinted from *U.S. Code of Federal Regulations,* Title 42, U.S. Public Health Service, Part 72.

- **a tertiary or outer shipping container.**

The space between the primary and secondary container must be filled with absorbent material sufficient to absorb the contents of the primary container should there be leakage during transit. The outside of the primary container should be examined and cleaned to remove blood, feces, or other contaminants before it is packaged for shipment.

The exteriors of packages containing cultures of, or suspensions of, etiologic agents should have affixed to them the "Etiologic Agent—Biomedical Materials" hazard warning label illustrated in Figure 3.1. The packaging and the labeling requirements of the regulation cited also apply to the local transport of etiologic agents and diagnostic specimens by courier or by other delivery services. Similar requirements and restrictions applicable to the shipment of etiologic agents, diagnostic specimens, and biological products by all modes of transportation (i.e., air, motor, rail, and water) are imposed by the Department of Transportation [131] and the U.S. Postal Service (*Postal Service Manual*), as well as by airline carriers and pilots' associations.

The importation of etiologic agents of human diseases, as well as their subsequent transfer within the United States, is regulated by the U.S. Public Health Service (USPHS) [128]. The U.S. Department of Agriculture (USDA) similarly regulates the importation and transfer of etioloic agents of plant and animal diseases [125]. In addition, the USDA Animal and Plant Health Inspection Service (APHIS)/ Veterinary Services (VS) Memorandum 593.1 establishes procedures for the "Importation of Cell Cultures Including Hybridomas." Examples of the appropriate USPHS and USDA application forms and permits are included in Appendix E.

A summary of the requirements of the federal agencies involved in the shipment of biological specimens has been published recently by the American Type Culture Collection (ATCC) (Rockville, MD 20852-1776). This document also describes the procedures used for packaging and shipping the different types of cultures of microorganisms and cells maintained by the ATCC [3].

Procedures for receiving and unpacking etiologic agents or other potentially infectious materials should be established by laboratories receiving these items. Often, such materials are received initially by shipping, clerical, or other nonlaboratory personnel. These employees should be given specific instructions to notify laboratory staff promptly of the arrival of such materials, and to deliver packages *unopened* directly to a designated area or person. Shipments of etiologic agents or diagnostic specimens should never be opened in offices or in shipping and other nonlaboratory areas.

In the laboratory, the designated specimen-receiving area should meet the facilities recommendation for Biosafety Level 2 [86,105]. Microbiological practices, including the wearing of laboratory coats, gloves, or other protective clothing, should be followed as applicable. A Class I or Class II biological safety cabinet provides the most suitable work station for opening packages and for the initial handling of incoming specimens [86] (see Chapter 3, Section J). Specimens that show any evidence of damage or leakage should be opened in a biological safety cabinet only by trained personnel wearing appropriate protective clothing.

Laboratories should have emergency contingency plans [147,148] for handling damaged shipments. Such plans are best prepared by the laboratory supervisor in conjunction with the laboratory staff and the safety officer. Emergency plans should be posted in a conspicuous place in the laboratory for immediate reference. Emergency plans should provide written procedures for dealing with

- **breakage or spillage of infectious materials,**
- **exposure of personnel to infectious materials by accidental injection, cuts, or other injuries,**
- **accidental ingestion or contact of mucous membranes with potentially hazardous material, and**
- **aerosols.**

Such emergency plans should include the following:

- **decontamination procedures,**
- **emergency services (whom to contact), and**
- **emergency equipment and its location.**

H. LABELING OF SPECIMENS WITHIN THE LABORATORY

Some form of labeling is necessary to maintain the identity of specimens in the laboratory and to ensure that the analytical results obtained are properly recorded and reported. In addition, it is the practice in many cases (and may be required as a condition for accreditation) that special hazard warning labels be affixed to specimens that are known to be hazardous (e.g., specimens obtained from patients known to be infected with hepatitis B virus (HBV) or human immunodeficiency virus (HIV), or from patients in high-risk groups for these infections, or when previous tests of the specimen have shown it to contain an etiologic agent).

The need for such special labeling is concerned more with ethical or regulatory issues (e.g., workers' right-to-know) than with laboratory safety. Unfortunately, the use of special hazard warning labels can inadvertently lead to the dangerous misconception that other clinical specimens, not so labeled, can be handled with less caution. Two levels of laboratory practice may thus evolve: one for handling hazard-labeled specimens and another for unlabeled samples. This must be scrupulously avoided: *all clinical material must be considered to be infectious, and must be handled with exactly the same precautions as are used for processing specimens with hazard warning labels.*

It is generally recognized that any clinical specimen may contain infectious agents (such as HBV or HIV) regardless of its source, the working clinical diagnosis, or the testing requested. For example, published reports indicate that from 1.0 to 1.5 percent of the adults in the United States have serological markers indicative of current or previous HBV infection [66]. Thus, even though the percentage of specimens containing an infectious agent may be higher among samples collected from hepatitis patients, the total number will probably be greater among routine specimens, which make up the vast majority of the materials received in most laboratories. The potential for worker exposure may, therefore, be actually greater from the more numerous routine specimens, which would not be identified with a hazard warning label. From the above considerations, it is clear that the "Universal Precautions" described by the CDC [34,38] must be followed when handling *all* clinical specimens, whether labeled or unlabeled.

I. PREVENTION OF AEROSOL AND DROPLET GENERATION

1. Introduction

Exposure to microorganisms dispersed or spread in the form of infectious aerosols or droplets is an important source of laboratory-acquired infection. Infectious aerosols may be composed of dry or liquid particles typically less than 5 microns in diameter, which can be produced during the course of many common laboratory processes. Such aerosols do not settle quickly and can be dispersed widely through a ventilation system or otherwise carried long distances by air streams. If inhaled, the particles in an aerosol are carried to the alveoli of the lungs. In contrast, droplets (particles typically larger than 5 microns in diameter) remain airborne only for a short period of time and are nonrespirable. Because of their mass, droplets tend to settle quickly on inanimate surfaces, or may be deposited on skin or mucous membranes of the upper respiratory tract. Accordingly, droplets pose risks of infection associated with direct or indirect contamination of the mucous membranes of the eyes, nose, or mouth as well as of skin, clothing, and laboratory equipment.

2. Control of Aerosols and Droplets

Almost any handling of liquids or of dry powders is likely to generate aerosols and droplets; certain operations such as pipetting, mixing, shaking, grinding, filtering, sonicating, flaming, and centrifuging have a high potential for aerosol production. Of these, pipetting may be the most important. Various reports indicate that pipettes are associated with many laboratory-acquired infections [98,99,100,102, 104,120]. Hazards relating to pipetting include the production of aerosols, aspiration of fluid into the mouth, and contamination of the mouthpiece by the operator's finger. The last two of these dangers can be avoided if mouth pipetting is strictly prohibited, as required by good laboratory practice. A wide variety of mechanical pipetting devices are available [63], and mouth pipetting under any circumstances is absolutely unacceptable.

To minimize aerosol production, pipettes should be drained gently with the tip against the inner wall of the receiving tube or vessel. No infectious material should be expelled forcibly from the pipette, and air should never be bubbled through a suspension of infectious agents in an open container. When handling organisms for which Biosafety Level 3 precautions are indicated (e.g., etiologic agents of tuberculosis, systemic mycoses, or Q fever), it is recommended that pipetting procedures be carried out in a biological safety cabinet. The equipment used in the other operations mentioned above should be selected for features designed to contain infectious liquids or aerosols. For example, blenders should have leakproof bearings and a tight-fitting gasketed lid. Blender bowls, tubes, and other devices likely to contain aerosols should be opened, filled, and emptied in a biological safety cabinet.

Centrifuges with sealed buckets, safety trunnion cups, or sealed heads are effective in preventing escape of liquids and aerosols (Figure 3.2). If fluid should escape from a cup or rotor during high-speed operation, the potential for extensive contamination

FIGURE 3.2 If a fluid containing an infectious agent were to escape from a centrifuge rotor or cup during high-speed operation, the potential for extensive contamination and multiple infections would be great. The use of sealed buckets, safety trunnion caps, or sealed heads is an effective means of preventing the escape of liquids or aerosols. Courtesy, John H. Richardson.

and multiple infections is great. For many specimens, however, such as urine, the standard clinical centrifuge is satisfactory. There have been comparatively few centrifuge accidents reported as the cause of laboratory-acquired disease, but some of these caused multiple infections because the accident created a large volume of infectious aerosol [141].

Instruments should be checked regularly to ensure that leakage does not occur during operational procedures. For ultracentrifuges, a HEPA filter should be installed between the chamber and the vacuum pump. If circumstances require such precautions, centrifuges and other laboratory instruments that can be enclosed and operated in specially designed safety cabinets are available. Only those instruments and cabinets intended for such a combined system should be used together, otherwise the expected containment may not be achieved. For example, the airstreams created by an ordinary benchtop centrifuge operating in the work space of an ordinary Class II biological safety cabinet can easily overwhelm the protective air curtain.

Sputum and other clinical specimens submitted for culture may contain unsuspected microorganisms, such as mycobacteria, which are highly infectious by the airborne route. Every effort should be made, therefore, to minimize the risk of their aerosolization. If generation of an aerosol is likely to occur during the processing of these specimens, the use of a biological safety cabinet is recommended strongly for these procedures.

Improper technique in the flaming of inoculating loops can result in the spread of infectious agents. Spatter and release of droplets or aerosols can be prevented by such methods as heating the shaft until the sample has been heat-dried before flaming the loop itself (Figure 3.3). Spatter can also be controlled effectively by using a side-arm burner or electric microincinerator. Flaming itself can be avoided by using sterile, disposable plastic loops.

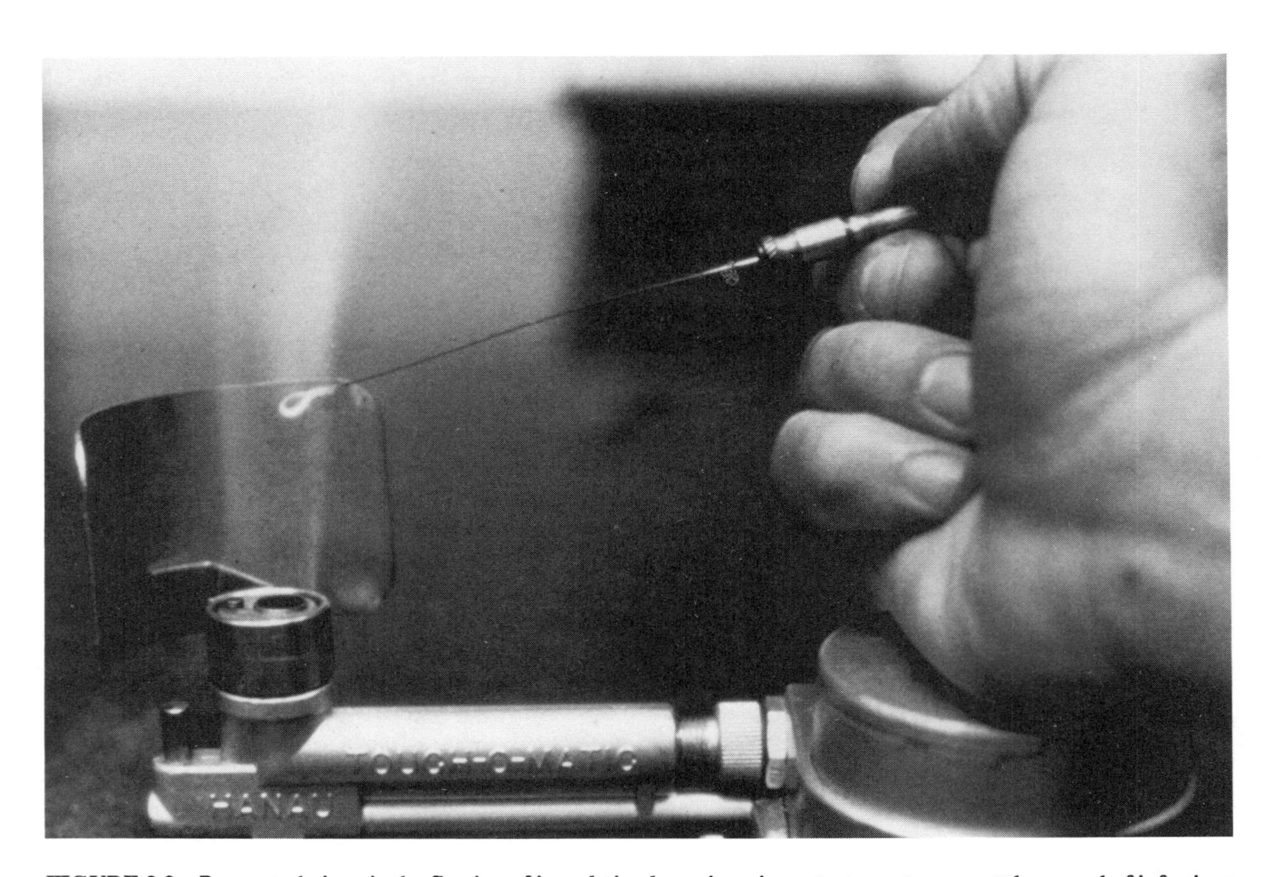

FIGURE 3.3 Proper technique in the flaming of inoculating loops is an important way to prevent the spread of infectious agents. Courtesy, National Institutes of Health.

Early models of certain laboratory instruments, such as cell sorters and other automated devices, were not designed for containment and may be a source of inadvertent contamination in the workplace. Later models generally have overcome the problem, but users are advised to test all equipment carefully in order to identify any biological hazards associated with its operation.

Regardless of the type of equipment used or the task performed, the objective is to prevent aerosol release and to avoid exposure of personnel. These ends can be accomplished by the laboratory practices described above and by the use of appropriate equipment, especially biological safety cabinets. Leaks or escape of aerosols can be detected by using an indicator such as fluorescein. It may be added to a sham specimen or to water, and processed with the system or procedure being tested. Its presence can then be determined on surfaces, on material collected from key locations, or in specimens from air samplers, by using an ultraviolet lamp for excitation. Other suitable methods may be devised.

J. CONTAINMENT EQUIPMENT

1. Introduction

The risk of exposure of laboratory personnel can be minimized by the use of carefully selected safety equipment. A primary objective of containment is to control aerosols, but in a broader sense safety equipment should serve effectively to isolate the worker from the toxic or infectious material being processed. In many situations, however, the need is just the reverse: i.e., to protect the product or the work from contamination originating with the worker or the environment. Finally, there is often the need to protect both the worker and the product, as in handling cell cultures and some clinical specimens, or in surgical procedures. The following examples are representative of the types of equipment designed to avoid the most common laboratory hazards, and these types of equipment are, for that reason, among the most important.

2. Biological Safety Cabinets

Most laboratory procedures generate aerosols that may spread infectious material in the work area and pose a risk of infection to the worker. Biological safety cabinets are used extensively to prevent the escape of aerosols or droplets and to protect materials from airborne contamination (Figure 3.4). There are three major types of this very useful safety device, referred to as Class I, Class II, and Class III. These instruments are distinct from horizontal or vertical laminar flow "clean benches," which should never be used for handling infectious, toxic, or sensitizing material. The Class I biological safety cabinet is an enclosure with an inward airflow through the front opening. It may be configured with a full-width open front or with an installed front closure panel to which arm-length rubber gloves may be attached. The exhaust air from the biological safety cabinet is passed through a HEPA filter so that the equipment provides protection for the worker and environment. The product in the cabinet, however,

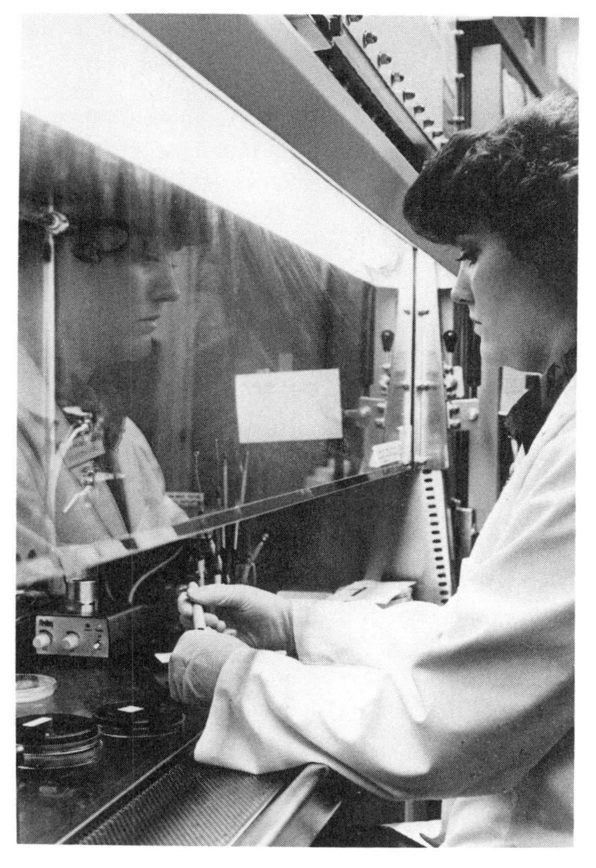

FIGURE 3.4 Biological safety cabinets, combined with protective gloves and laboratory coats, provide effective isolation of the worker from the toxic or infectious material being handled. Courtesy, John H. Richardson.

is subject to contamination by organisms that may be present in the air supply.

Class II biological safety cabinets provide protection to the worker, the environment, and the product. The airflow velocity at the face of the work opening is at least 75 linear feet per minute (lfpm), and both the supply and the exhaust air are HEPA-filtered. Class I and Class II cabinets are partial containment devices, which, if used in conjunction with good laboratory practices, can dramatically reduce the risk of exposure of operators to infectious aerosols and droplets.

Figure 3.5 [73] shows the airflow patterns and operating velocities for the five types of Class I and Class II biological safety cabinets produced in the United States. All of these biological safety cabinets provide a comparable level of protection for the user against exposure to infectious aerosols and droplets, in that the velocity of the protective inward airflow through the work opening is essentially the same. The air quality within the Class I cabinets reflects that of the laboratory room from which it is drawn, since there is no filtration of the supply air. The Class II types provide a very high quality, low-particulate or particulate-free atmosphere within the work chamber. Class IIA cabinets are generally suitable for procedures involving clinical specimens, and thus are the most commonly used biological safety cabinet.

It is emphasized that biological safety cabinets are not chemical fume hoods. Some of the air (30 to 70 percent) drawn in through the work opening of Class IIA, IIB1, and IIB3 cabinets is recirculated within the cabinet. Accordingly, users should be aware of the possible buildup of hazardous concentrations within the cabinet if toxic, flammable, or explosive materials are used. In addition, users of Class IIA cabinets should know that nonparticulate toxic, flammable, or explosive materials are not removed by HEPA filters, and are thus discharged back into the laboratory room.

Class IIB3 units are functionally the same as those of Class IIA except that the exhaust air from the former is ducted to the outside directly or via a nonrecirculating exhaust system rather than back into the laboratory room. Class IIBl and IIB2 cabinets exhaust 70 percent and 100 percent, respectively, of the intake air and provide containment of infectious aerosols. Those contemplating the purchase of Class IIBl or IIB2 cabinets should be aware of their high air demand (700 to 1200 ft^3/min), increased energy requirements, and higher purchase and operating costs.

The Class III cabinet is a totally enclosed, gastight work space equipped with protective gloves. It is ventilated with HEPA-filtered air and operated with a negative air pressure of at least 0.5 inches of water in the cabinet work space. The exhaust air is passed through two HEPA filters, installed in series, before being discharged to the outside of the building, usually through a dedicated exhaust system. Class III cabinets provide the highest level of worker, product, and environmental protection and are appropriate for work with exotic high-risk biological agents, including those in the Biosafety Level 4 category.

The operational efficiency of each biological safety cabinet should be specifically tested and the system certified before the instrument is placed in operation after installation, and subsequently on an annual basis. Recertification is also required if the unit is relocated or if maintenance that may affect performance is done. Maintenance work on biological safety cabinets should be performed by trained service personnel only (see Chapter 5, Section B). In addition, cabinet users should understand the operation of the equipment, its limitations, and the proper procedures to be followed. Laboratory directors are responsible for providing such training.

3. Pipetting Devices

Pipettes are among the most commonly used pieces of equipment in the biomedical laboratory, and their misuse has been related to a significant number of laboratory-acquired infections [100]. Regrettably, many laboratory workers were taught to pipette by mouth, even after the associated hazards were recognized. These individuals should be required to give up the old practice and learn to use the pipetting aids that are now available for any application [63] (Figure 3.6). The importance of these aids cannot be overemphasized, and any device requiring mouth suction should be considered unsafe and inappropriate for use in the biological laboratory. Mouth pipetting of any material under any circumstances should be explicitly prohibited.

4. Sonicators, Homogenizers, and Mixers

Operation of these or similar instruments may create hazardous aerosols and lead to exposure of

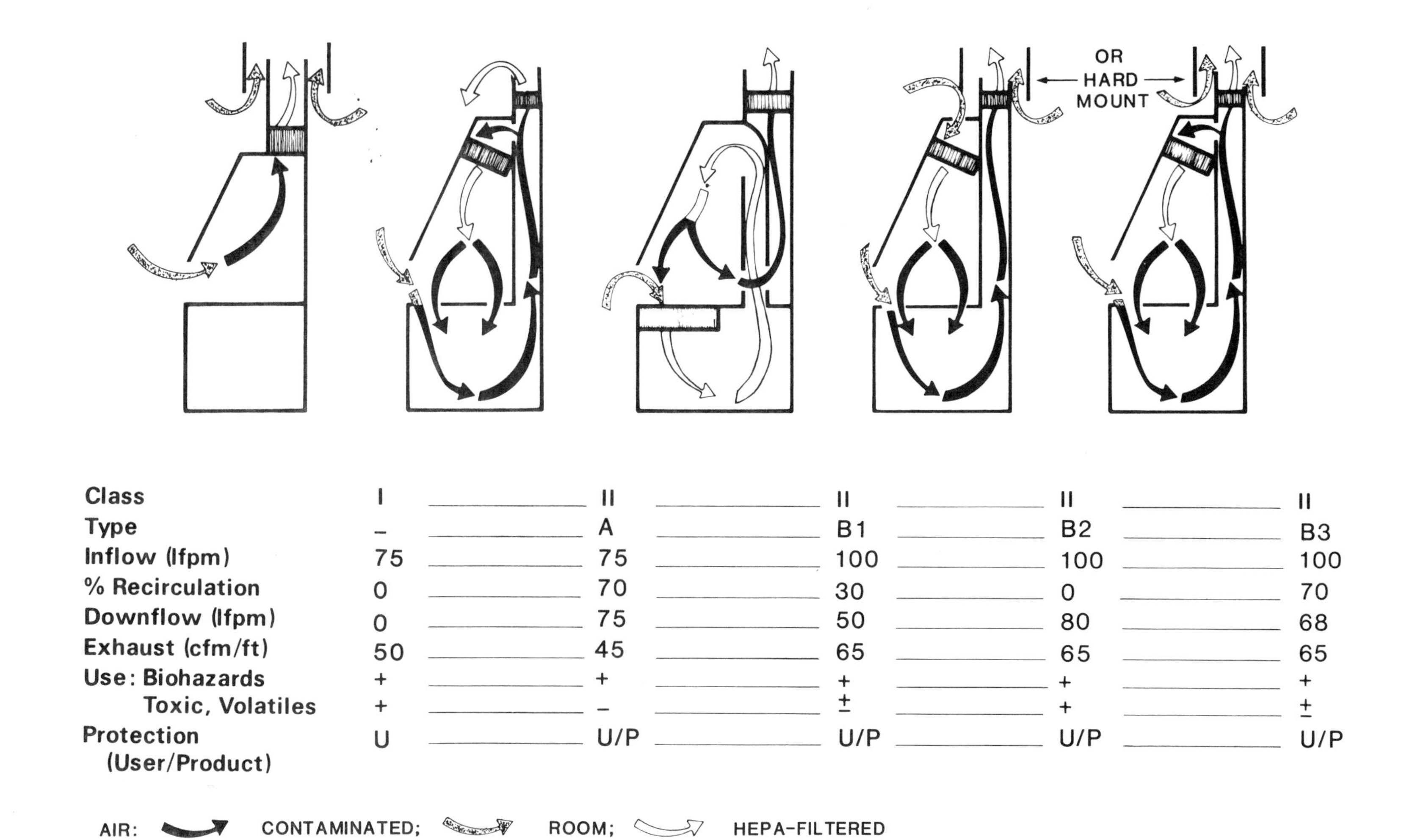

Class	I	II	II	II	II
Type	–	A	B1	B2	B3
Inflow (lfpm)	75	75	100	100	100
% Recirculation	0	70	30	0	70
Downflow (lfpm)	0	75	50	80	68
Exhaust (cfm/ft)	50	45	65	65	65
Use: Biohazards	+	+	+	+	+
Toxic, Volatiles	+	–	±	+	±
Protection (User/Product)	U	U/P	U/P	U/P	U/P

AIR: CONTAMINATED; ROOM; HEPA-FILTERED

FIGURE 3.5 Airflow characteristics of Class I (negative pressure) and Class II (vertical laminar flow) biological safety cabinets. Adapted from National Sanitation Foundation Standard 49 (revised May 1983) by G.P. Kubica, Centers for Disease Control.

personnel unless extreme caution is exercised. If indicated by the characteristics of the material being processed or the agents involved, the instrument should be operated in a biological safety cabinet. Blenders should be designed to prevent leakage from the rotor bearing or at the cover. Caps and gaskets should be in good condition and the system checked to ensure that leakage does not occur during operation. When the risk of exposure to infectious aerosols is present, blender bowls, tubes, and other containers should be opened in a biological safety cabinet.

5. Clothing, Masks, and Face Shields

Laboratory coats, gowns, safety glasses, face shields, masks, and gloves offer some personal protection and are often used in combination with other safety devices such as biological safety cabinets. Special laboratory clothing protects street wear from contamination. It should not be worn outside of the laboratory. Each of these items has a particular use in protecting the worker and should be used when circumstances require. Gloves are especially important when handling any potentially infectious material such as blood or other biological specimens. Safety glasses, face shields, and masks may protect mucous membranes of the eye, nose, and mouth from splash or droplet hazards during operations performed outside of a biological safety cabinet.

K. BIOSAFETY IN LARGE-SCALE PRODUCTION

1. Introduction

Microbial cultures of greater than 10 liters in volume (defined as large-scale [136]) present biosafety concerns very similar to those described for small-scale (<10 liters) experiments in the laboratory. All the recommendations described in the CDC/NIH guidelines (Appendix A) for laboratory-scale research should, therefore, be followed for large-scale production, with the addition of the recommendations described below. These recommendations are equally applicable to cultures as small as 20 liters and as large as 10,000 liters.

This section addresses only issues of biological safety relative to infectious agents and does not deal with other major areas of potential risk in large-scale

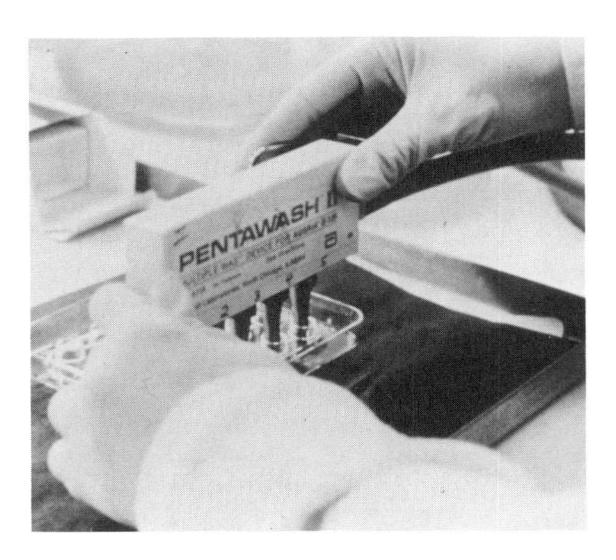

FIGURE 3.6 The wide variety of available pipetting aids make mouth pipetting an unnecessary and obsolete practice. Courtesy, John H. Richardson and National Institutes of Health. (Figure continued on next page.)

production (e.g., end-products, by-products, media components, and nonviable biological agents). These subjects exceed the scope of this book, but other sources [118], and references contained therein, provide pertinent information. Similarly, this section does not address the biological safety issues that pertain to the large-scale growth of mammalian cell cultures. This topic is discussed briefly in Section D of this chapter.

The physical containment typically implemented for large-scale production serves to protect the worker. The self-contained design used for large-scale production should essentially eliminate the generation of aerosols, one of the most common causes of laboratory infections [100]. Standard operating procedures (SOPs)—including validation of equipment's function, biological disposal, and specific methods of operation—should be written and closely followed for all large-scale productions. Implementation of these protocols should facilitate the maintenance of a safer environment in which to work and produce the highest quality of resultant product.

Special circumstances inherent in large-scale production may actually reduce risks to levels lower than those encountered in laboratory research. Potential biological hazards associated with the specific organisms used for large-scale production will typically have been defined and assessed in the research

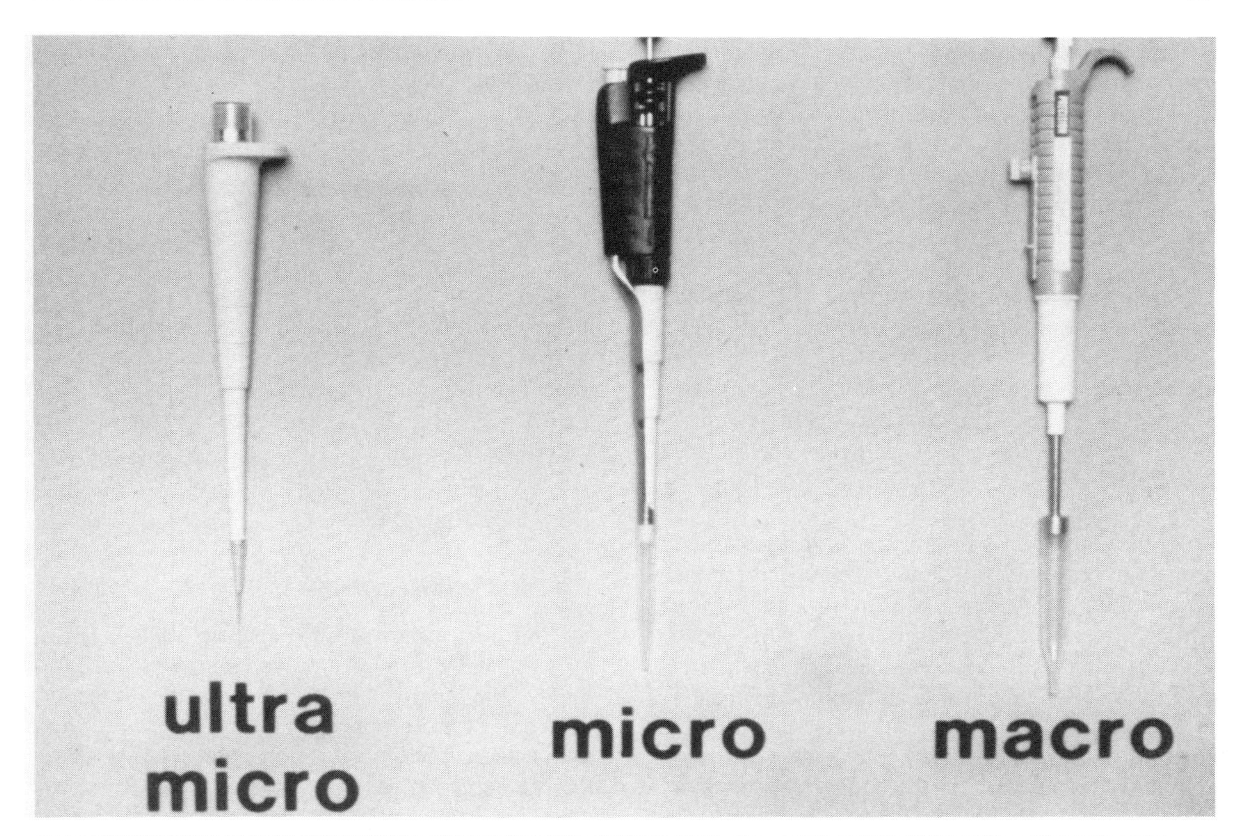
ultra
micro
micro
macro

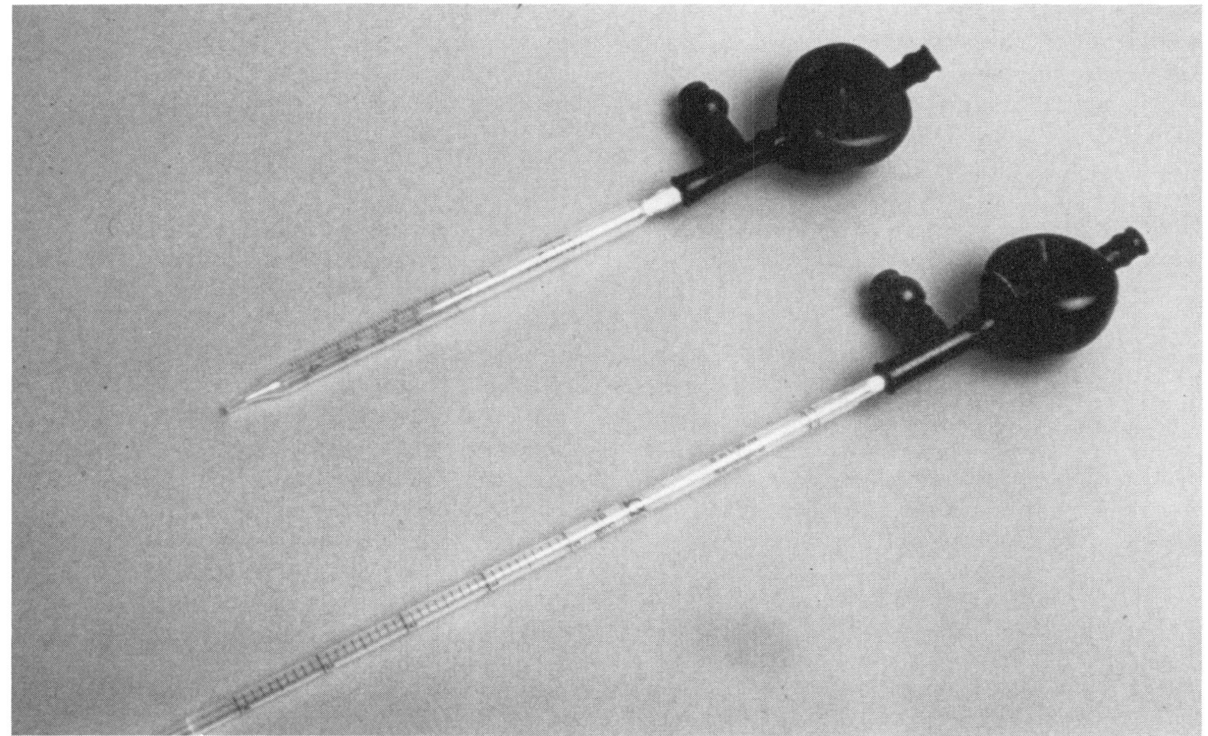

and developmental stages before large-scale production is initiated. Because fewer biological unknowns or variables are present, the biological hazards should be reduced. This situation is especially true when comparing large-scale culture of well-characterized microorganisms with laboratory-scale culture of clinical isolates or of soil isolates, where a tremendous diversity of unknown organisms is present. When the proper precautions are implemented and maintained, large quantities of potentially hazardous microorganisms may be grown safely [65,135]. Many of the organisms or cell lines of interest for production purposes (e.g., Bacillus and yeast) have had an extended industrial history of safety on the basis of which risks can be accurately and confidently evaluated.

Where large-scale production is initiated in the process of commercialization of a desired product, the constraints enforced to ensure product integrity typically result in increased safety to personnel as well. As numerous laboratories have developed large-scale production facilities for the commercialization of biological products, the need for carefully designed and implemented procedures has become increasingly important for the safety of laboratory workers, the community, and the environment. The following procedures are intended to allow handling of suspected or known hazardous organisms at large-scale levels with proper concern for health and safety. The recommendations are not intended to apply rigidly to all situations. As experience is gained in scale-up operations, more or less stringent guidelines for individual operations may be indicated by the safety department, biological safety officer, or biosafety committee. The procedures reflect the best judgment of acceptable techniques and applicable legal requirements [4,13,18,51,82, 105,124,129,130, 132,133,134,136]. The recommendations rely heavily on the NIH "Guidelines for Research Involving Recombinant DNA Molecules" [136], which have proved dependable as a foundation for laboratory safety programs whether or not recombinant DNA is involved. These recommendations are broadly applicable because they are based on the potential pathogenicity or infectivity for the host. These procedures incorporate safety concepts and guidelines published in documents of the National Institutes of Health [133,134,136], the Centers for Disease Control, in conjunction with the National Institutes of Health [105], the U.S. Department of Agriculture [124], the American Industrial Hygiene Association [4], and the Medical Research Council of Canada [82].

2. Organization and Responsibilities

Institutions planning to scale up the production of biotechnology-related products should have, at least, a safety department, a biological safety officer, and a biosafety committee, with the responsibilities described in Chapter 5. Scientists, technicians, equipment workers, and maintenance and custodial personnel with access to the large-scale production area should all be considered candidates for medical surveillance, depending upon the organism being grown and the product produced.

3. Containment

Containment levels for organisms such as fungi, bacteria, and viruses are grouped into classes according to their perceived or potential hazard to humans. Selection of an appropriate biosafety level for work with a particular agent depends upon such factors as the virulence, pathogenicity, biological stability, mode of transmission, endemic nature, and communicability of the agent; the function of the laboratory; the procedures and manipulations of the agent; the availability of effective vaccines or therapeutic measures; and the quantity and concentration of the agent [105]. The NIH "Guidelines for Research Involving Recombinant DNA Molecules" [136] considers cultures greater than 10 liters in volume to be large-scale and, for specific organisms, to require higher levels of physical containment than cultures less than 10 liters in volume. Increased containment for large-scale culture is typically reflected in the requirement for more extensive physical design, and not for operating at a higher biosafety level.

The risks associated with the use of microbial cultures are controllable through two containment approaches. For some work, an appropriate approach is biological containment: the use of species or strains that pose a reduced risk to workers or to the environment. The more widely applicable approach, however, is physical containment: a combination of laboratory practices, design, and equipment. Physical containment, through the implementation of good laboratory practices and proper monitoring, certification, and maintenance of the facility, leads to a significantly safer and more productive work environment.

A National Institute for Occupational Safety and Health (NIOSH) survey of six biotechnology companies [51] showed that companies with many years of experience in fermentation technology tended to emphasize more sophisticated, more effective, and safer practices for handling infectious agents than newly established companies.

All facilities and equipment used to provide containment should be tested before initiation of a program and periodically thereafter by trained personnel. Such monitoring should include checks of room air balance, biological safety cabinets, supply and exhaust filters, sterilizers, and centrifuges. Laboratory personnel should ensure that biological safety cabinets are appropriately certified prior to use. A systematic, scheduled program of preventive maintenance should be implemented for agitator seals, control valves, pressure relief valves, and equipment and facility safeguards [51]. Sterilization of fermentors, feed lines, feed tanks, inoculating devices, exhaust ports, sampling ports, and extraction devices should be validated routinely.

Maintaining, testing, or cleaning of facilities or equipment should not be allowed until all surfaces needing servicing have been decontaminated. Spills should be disinfected by laboratory personnel before housekeepers are allowed to give assistance.

Typically, large-scale production facilities are specifically engineered for maximal physical containment to ensure both personnel safety and product integrity. The specific minimum requirements for physical containment are described in the NIH "Guidelines for Research Involving Recombinant DNA Molecules" [136]. The large-scale containment classifications, BSL1-LS, BSL2-LS, and BSL3-LS (Biosafety Levels 1, 2, and 3, large-scale, respectively) are required for organisms for which the corresponding containment levels BSL1, BSL2, and BSL3 (Biosafety Levels 1, 2, and 3, respectively) are required for small-scale (<10 liters) research. No provisions are made for large-scale growth of microorganisms that require the Biosafety Level 4 containment at the laboratory scale. (Consult Appendix A for the small-scale containment levels for specific microorganisms.) The appropriate implementation of safeguards and of protective engineering controls at the time of the design of the facility or laboratory can reduce human exposure and avoid expensive retrofitting. Specific details of engineering and design features that reduce the potential biological risks associated with large-scale microbial production can be found in the NIH guidelines [136].

4. Inactivation

Vessels used for large-scale production are typically steam sterilized in place, in the absence or presence of medium, prior to fermentation. Temperature-sensitive tapes, crayons, or, preferably, thermocouples, may be used to monitor sterilization procedures. The growth medium is either sterilized in the system or sterile medium is introduced via a closed system designed to maintain physical containment. Connections should be sterilizable in place. Metal containers, designed to preclude escape or entry of viable organisms during inoculation, should be used. These inoculation vessels should be sterilizable after the transfer process, and prior to their removal from the production vessel, to minimize release.

Following growth, cultures in the production vessel should be inactivated either chemically or thermally prior to initiating the product recovery processes. Inactivation is used in this context to refer to the reduction in the total number of viable target microorganisms in a large-scale culture to a number comparable to or less than that obtained in a small-scale laboratory culture (<10 liters). This reduction enables a culture that is grown in a large-scale, closed system to be handled according to the containment level applicable for the corresponding laboratory-scale culture. Inactivation, in this context, is therefore an interim action that is to be followed by decontamination prior to the disposal of microbial cultures. As an alternative to inactivating cultures in the production vessel, cultures can be inactivated by cell rupture or by further steps in subsequent stages of the process, if these are designed into the same closed system. When the viable biological agent itself is the desired end-product (e.g., some vaccines), the culture would not be inactivated but would be maintained within a closed system to avoid human exposure. The specific method of inactivation (see Chapter 4) depends on the organism or cell culture employed and upon the product to be isolated. Inactivation efficiencies should be validated as described in the SOPs. Physical containment during processing steps following culture inactivation (i.e., steps involved in the recovery and purification of the defined product) is not required but is recommended when possible and feasible. All liquid and solid biological waste should be decontaminated chemically or thermally prior to disposal. In-line thermocouples are especially effective in validating the thermal inactivation of large quantities of liquid cultures prior to disposal. If other primary equipment, such as centrifuges, is used in-line for harvesting cells prior to inactivation, this equipment should also be decontaminated appropriately.

Water supplies or storage tanks provide ideal entry points for biological contaminants. Treatment of water with ultraviolet light, ozone, or passage through specifically selected filters significantly reduces the potential for its contamination. The quality and type of water processing should be specified in the SOPs.

5. Disposal

All biological waste generated through routine procedures or as the result of accidents should be decontaminated prior to its disposal, and should be segregated from nonbiological waste and from radioactive waste. Solid trash and small volumes of waste should be disposed of by using the procedures described for laboratory research. Contents of the production vessel should be physically contained (e.g., with dikes or decontamination tanks) to confine spills and leaks and to allow for their rapid and efficient decontamination. All liquid collected by these procedures should be properly decontaminated by using validated procedures to prevent the release of viable organisms or cells into the environment. The contents and all associated materials and equipment should be inactivated either chemically or physically prior to disposal. If decontamination tanks are used, the method of inactivation should be validated for each organism employed. Once inactivated, even large quantities of liquid culture can be disposed of by discharge to the sewer lines, provided that such action is permitted by the relevant state and local agencies. Air discharged from fermentors should be filtered through a HEPA filter, incinerated, or otherwise treated prior to release.

Personnel involved in the cleanup of accidentally spilled waste should proceed as described in Chapter 5 for spills with laboratory-scale cultures. Particular precautions should be taken to handle and decontaminate these large quantities of culture, as well as the absorption materials used for cleanup.

6. Exposure

The self-contained design of the large-scale production system, the inoculation method, and the in-place inactivation of the vessel contents following production essentially eliminate the release of aerosols, thereby reducing human exposure. Nonetheless, air and surfaces should be monitored to validate the integrity of the systems during the fermentation or processing procedures. Surfaces can be monitored for microbial release by any of five basic methods: rinse, swab-rinse, agar contact, direct surface agar plating, or vacuum probe surface method [4]. Other procedures are modifications of these basic methods. In a survey of six genetic engineering companies, agar contact was found to be the method most frequently used [51]. Organism detecting and counting (RODAC) plates are commercially available with a variety of media and are used routinely. The method works well and rapidly as a qualitative procedure.

Numerous methods are available to sample airborne microorganisms [4]. The more frequently used methods include settling plates, air impingers, and filters. Settling plates provide qualitative information and consist simply of an open petri dish containing appropriate culture medium onto which particles settle due to gravity. Air impingers are intrinsically more quantitative because they pull air at a fixed rate and volume onto the surface of the medium. Large and defined quantities of airborne microorganisms can be concentrated and analyzed. The most common method for sampling air is filtration. A fixed volume of air can be passed through a filter of a selected pore size. Particles and organisms can be flushed from the filter, and the filtrate analyzed on an appropriate medium, or the filter itself can be overlaid directly with an appropriate medium. The use of filters is limited, because of dehydrating effects, to the detection of spores and of resistant vegetative cells. Selection of an appropriate sampling method depends upon the nature of the particles of interest, their expected concentration, and the need for quantification. The reader can consult Table XVIII in reference 6 for a more detailed description of these and alternative sampling methods.

Surface and air sampling methods should be selected to detect biological contaminants, as well as the production organism itself. Adventitious agents expected in large-scale production are similar to those found in laboratory research (e.g., viruses, bacteria, and fungi). Potential candidates depend upon the specific organism under study and the medium employed.

7. Conclusion

Large-scale microbial culture operations can be done safely, despite the risks associated with some microorganisms. Personnel can function safely and efficiently at any scale by using the proper facilities and equipment.

L. BIOSAFETY IN PHYSICIANS' OFFICE LABORATORIES AND OTHER SMALL-VOLUME CLINICAL LABORATORIES

Each small laboratory, including those in physicians' offices, should have a safety program even if the laboratory has only one employee (see Chapter 5). For those employees without training in the biological sciences, special effort should be made to provide simple but effective instruction. The program should be tailored to the laboratory function, with proper emphasis on providing training in the seven basic rules of good microbiological practices as outlined in Section F of this chapter. Indeed, all of the general principles outlined in this chapter for the safe handling of infectious agents apply to small-volume laboratories as well as large-volume ones.

Many of these small-volume laboratories may be handling human blood or blood components that could be infectious. If this is the case, the standard biological practices as well as the special practices recommended by the CDC/NIH guidelines for Biosafety Level 2 should be used (see Appendix A), and the Universal Precautions described in Appendix C should be followed. Prudence should also be exercised to minimize exposure to toxic chemicals and radionuclides.

Special precautions should be taken to ensure that waste is managed in a safe, responsible manner. Potentially infectious waste should preferably be decontaminated on site. Liquid waste, which may be contaminated with an infectious agent, can be steam autoclaved or decontaminated by using a chemical disinfectant that is effective for the intended use. Decontaminated liquids can than be poured carefully down a drain connected to the sanitary sewer. Solid waste, including contaminated "sharps" (e.g., hypodermic needles and broken glassware), should be packaged in sealed, leak- and puncture-resistant containers for transport and disposal. Such properly contained waste presents minimal risk to waste handlers and can be safely managed as municipal solid waste. If this practice is prohibited by local regulations, then the prevailing practices should be followed. Human excreta should be disposed of through the sanitary sewer. The practices just described are appropriate for the small laboratory that is involved in testing patients' specimens. The safe disposal methods discussed in Chapter 4 should be followed in the small biomedical laboratory where infectious agents are isolated and grown in cultures.

Further information on safety in the office laboratory may be found in references 12 and 54.

4

Safe Disposal of Infectious Laboratory Waste

A. INTRODUCTION

Human activities produce biological waste in the form of human excreta or other discarded materials, much of which may contain infectious microorganisms. Such waste, if untreated, has varying degrees of potential to cause disease. Existing methods of sanitation have served effectively to protect the public's health from any disease associated with biological waste. Our understanding of the conditions required to prevent the transmission of disease has allowed the development of simple, yet highly effective management techniques for handling biologically contaminated waste. A brief review of some of the conventional measures used to protect the public's health follows.

Historically, human excreta have been linked to outbreaks of disease such as dysentery, poliomyelitis, typhoid, and cholera. Such waste may contain high concentrations of pathogens that can contaminate food and water supplies. To minimize the opportunity for such cross-contamination, several fundamental principles of sanitation are applied. The first principle consists of providing a physical barrier. Sanitary sewer systems, which consist of pipes and pumps to convey pathogen-laden sewage to a treatment plant, effectively provide such a barrier. Once the waste reaches the sewage treatment plant, other mechanisms help to reduce the disease-causing potential of the material.

The treatment of sewage usually consists of biological degradation of the organic material. The physical and biochemical conditions that are optimal for such degradation are often hostile to the survival of many human pathogens. The end result is a significant reduction in the numbers of viable pathogenic microorganisms remaining in the treated waste.

After the organic load in the sewage has been degraded to the desired level, the effluent from the sewage treatment plant is usually subjected to a final disinfection step before being released into the environment. This step is accomplished by chlorination/dechlorination, exposure to ultraviolet light, ozonation, or some other procedure. The process ensures that the concentration of pathogens in the effluent is reduced to an acceptable level.

Like sewage, much solid waste or refuse produced by man is contaminated with biological agents capable of causing infection in man [39], and many of the same control mechanisms apply. For both storage and transport for disposal, the barrier system is again particularly important for the protection of individuals who must handle the waste. The plastic refuse storage bag, the dumpster, and the enclosed refuse-handling vehicle all provide barriers to minimize the potential of the waste to contaminate the environment.

Most solid waste is either disposed of directly in sanitary landfills or treated first by incineration to reduce its volume. Sanitary landfilling is a controlled disposal method designed largely to protect the public's health and the environment and consequently has largely replaced the open dump. Besides producing smoke and odors, open dumps provide ideal habitats for the propagation of disease-carrying vectors of concern to man, such as rats, flies, and mosquitoes. The sanitary landfill eliminates this habitat by compacting the waste and providing a daily earth cover, also compacted, to seal off the waste from the general environment. The earth above and below is the barrier. Because the conditions of biodegradation within the landfill are hostile to many human pathogens, their numbers decrease with time [39,116].

It is recognized that sanitary landfills may produce liquid leachate that can carry viable microorganisms to the earth underlying the landfill. As a leachate percolates through the earth, remaining pathogenic microorganisms are reduced further in concentration by hostile environmental factors and by soil filtration. This principle is used to advantage in the leach fields of home septic systems [87].

Incineration of municipal waste is done primarily to reduce the volume of the waste. Because municipal waste is generated by both healthy and sick individuals, it contains the same array of human pathogens as those associated with hospital waste. Although it is not their primary goal, well-designed and well-operated municipal incinerators can provide effective destruction of pathogens in the same way that a hospital incinerator does. Municipal incinerators often operate at higher temperatures and with longer gas retention times, thereby enhancing their effectiveness for the destruction of pathogens.

In summary, the application of long-standing principles of sanitation is effective in controlling the threat to the general public's health associated with biological waste. In general, extraordinary pathogen control measures have not been shown to be necessary except where the physical nature of the waste (e.g., contaminated "sharps" that penetrate barriers) presents problems directly to the waste handlers, or the waste itself has an exceptional bioload in a mobile form such as a liquid culture of a pathogenic agent. In such cases special waste packaging may be needed, or on-site decontamination applied, to reduce the concentration of pathogens to acceptable levels.

B. INFECTIOUS POTENTIAL OF LABORATORY WASTE

For laboratory waste to cause infection, six essential factors must be present. These factors are as follows:

1. The presence of an infectious agent that is capable of invading and multiplying within a human host.
2. An environment for the infectious agent that functions as a reservoir, allowing it to survive and, perhaps, to multiply.
3. A mechanism for the agent to escape from the reservoir.
4. A mode of transmission from the reservoir to a human host.
5. A means for the agent to invade, penetrate, or enter a human host.
6. A human host that is susceptible to infection by the agent.

The absence of any one of these factors will interrupt the infectious process and human disease will not ensue.

An understanding of these factors is necessary for assessing public health risks and the risk of occupationally acquired illness that may be associated with the management of infectious waste. Infectious waste implies the presence of viable pathogenic microorganisms in sufficient concentration to infect a susceptible human host. There is no risk of disease when the concentration of pathogenic organisms is below that which is capable of invading the host and multiplying within it. The source of the infectious waste may be a patient in a health care facility, an experimental animal in an infectious disease vivarium, or the culture medium used for the propagation of an infectious agent. The processes that generate the waste provide the means by which the infectious agent escapes from the reservoir. Thus the agent, reservoir, and means of escape will always be present in institutions that generate infectious waste. Treatment and disposal strategies that protect the public's health and prevent occupationally associated infection will therefore be necessary to block transmission of the agent and exposure of a susceptible host.

1. Risks to the General Public's Health

Risks to the general public's health can be associated only with indirect modes of transmission, because the public is not directly exposed to the institutional reservoirs or the infectious waste generated by them. For indirect transmission to occur, the infectious agent must be capable of survival outside of the reservoir for an extended period of time. There also must be an opportunity for a susceptible host to be exposed to the agent. Modern sanitation practice, as discussed in the introduction, minimizes the occurrence of such events. A properly functioning community sanitary landfill, solid waste incinerator, or municipal sewage treatment facility provides adequate containment and treatment for infectious waste, even when the waste is introduced without prior treatment.

2. Occupational Risks

Unlike the general public, workers who generate, handle, and process infectious waste have the potential for direct exposure to infectious agents. Exposures can occur through direct inoculation, such as when a worker is cut accidentally by a piece of contaminated glass, or through inhalation when the handling process generates aerosols. Occupationally acquired illness associated with the handling of infectious waste has been reported [59]. Protection against occupationally acquired illness is achieved through appropriate waste handling and treatment methods, which either contain the waste or inactivate the infectious agent. This chapter provides guidance to assist institutions and generators in establishing prudent practices for the management of infectious waste.

C. CHARACTERISTICS OF INFECTIOUS LABORATORY WASTE

Infectious laboratory waste is characterized principally as waste that contains microorganisms capable of causing infection in a healthy, susceptible host. Hospitals, health care facilities, medical research institutions, and industrial facilities can generate infectious laboratory waste. Categories of operation that produce infectious waste include the following:

- **operations that involve the processing and analysis of specimens for diagnosis, separation or purification of cells or substances from human blood and body fluids, and in vitro and in vivo methods for the propagation of pathogenic microorganisms;**
- **medical operations in which invasive procedures are likely to result in waste contaminated with blood and body fluids from an individual who harbors an infectious agent;**
- **veterinary operations involving the study of zoonotic disease in which infected animal carcasses and tissues, contaminated fomites such as disposable instruments and supplies, and contaminated bedding materials are produced;**
- **anatomical pathology services where workers process specimens from individuals either known to harbor, or who are at an increased risk of harboring, an infectious agent;**
- **diagnostic, research, and industrial operations that involve the collection and processing of bulk quantities of human blood, blood derivatives, or body fluids; and**
- **the production of biological products in which pathogenic microorganisms are used, such as vaccines.**

Biological waste with objectionable or putrescent characteristics, containing viable microorganisms that are either not known to be hazardous to humans or are minimally potentially hazardous, is not considered infectious laboratory waste. Examples include tissues or medical waste generated from the care of individuals who have not contracted an infectious disease; solid waste including such items as soiled diapers, animal bedding materials or pet litter, animal carcasses, and garbage from food processing plants and eating establishments. Objectionable nonhazardous medical waste is typically generated in extended care facilities and ambulatory health care services. Adherence to good personal hygiene and prudent sanitation practice affords adequate protection to individuals involved in the handling and disposal of this type of waste.

D. RESPONSIBILITY FOR THE SAFE HANDLING AND DISPOSAL OF INFECTIOUS WASTE

The primary responsibility for the safe handling and disposal of infectious waste resides with the generator of the waste. This responsibility extends to the ultimate point of disposal even when there are other parties involved in handling the waste. The generator should conduct inspections or take other measures to ensure that the waste is being handled and disposed of properly, even though management of infectious waste is also the concern of waste haulers and treatment facility operators. In addition, there may be federal, state, or local regulations controlling medical waste disposal and recordkeeping that must be observed.

The major problem associated with infectious waste is the potential for occupational exposure. The disposal of infectious waste should, therefore, be performed in an effective manner that minimizes the potential for exposure of those who, by virtue of their employment, must handle the material.

It is incumbent upon the scientific community to educate the general public about the effectiveness of current sanitation practices in protecting the public's health, and to direct legislators' attention to the problems of the occupational hazards associated with the handling of infectious waste.

1. Generators of Infectious Waste

The initial recognition of a potential hazard should set in motion all of the mechanisms from source to final disposal. Recognition is made most effectively by the generator of the waste. The responsibility for proper handling, treatment, and disposal of infectious waste materials resides with all those who knowingly come in contact with these materials. However, the major responsibility for proper handling belongs to the initial generator of potentially hazardous materials, because such waste material should be identified and segregated according to its degree of potential hazard. Once the material has been identified as infectious waste, proper packaging and containment should be ensured until decontamination or inactivation can be accomplished.

The most senior official of the facility generating the waste has the responsibility for the development of a waste management program. Such a program should ensure proper containment of infectious waste and the development and implementation of appropriate methods for the efficacious decontamination of this waste. It is incumbent upon management to ensure that proper operational controls of selected treatment methods are maintained.

No waste management program is functional unless all appropriate personnel are cognizant of the aims of the program and trained in the procedures for handling the waste. Management should provide resources and ensure that training programs are developed and implemented. Training should be a continuing process.

2. Haulers and Waste Treatment Facilities

Infectious waste is often decontaminated at the generating facility prior to its transport to a disposal site. Decontamination protects the waste hauler from the risk of infection.

In those instances where the waste hauling company transports untreated infectious waste to a treatment and disposal site, adequate physical containment measures should be provided to minimize occupational exposure. The processing of infectious waste by a treatment facility requires the same stringent attention to detail that is required of the generator treating its own waste. The treatment company has the responsibility for ensuring that all procedures are adequate and that all systems are functioning correctly. Both hauling and treatment companies should ensure that all of their personnel are made aware of the potential hazards of exposure. Personnel should be properly trained in all of the pertinent aspects of the containment, the handling, and the treatment of infectious waste. Contingency plans should be developed to handle accidents that may occur.

E. WASTE HANDLING AND TREATMENT METHODS

The prudent management of infectious laboratory waste requires the development of site-specific plans. Procedures developed by personnel within a facility will be appropriate for the specific needs of that facility and may gain a higher level of acceptance than will procedures imposed from outside sources. The process of developing a waste management plan is, in itself, acknowledgment of the need to accept responsibility for laboratory waste.

1. Basic Principles

Persons who generate infectious laboratory waste are responsible for preparing the waste so that potential occupational exposures and environmental contamination are minimized. Infectious waste needs to be segregated by the generator from other waste streams. This process will obviate the need for decision-making by support services personnel. The waste can then be treated on-site to reduce the concentration of the pathogen to an acceptable level (decontamination), or packaged in a way to prevent subsequent exposure of other persons having to handle the waste prior to terminal treatment. Packages of infectious waste need to be identified so that the potential hazard clearly can be recognized and understood by others. The universal biohazard symbol is used for this purpose.

2. Containment

A variety of packaging items for containment and transport of infectious waste are available. Infectious waste containers serve as primary barriers to protect the worker and to minimize the chance of environmental contamination (Figure 4.1). Typically, these containers are made from leak-resistant paper or cardboard, stainless steel, or temperature-resistant polymers. The nature and volume of the waste, the terminal treatment method, and their costs are principal factors to consider in the selection of the mode of packaging.

Solid waste can be packaged safely in sturdy bags or boxes. Flat trays with sealable lids are suitable for containing pipettes and other laboratory supplies during decontamination. Bulk liquids may be collected in leak-proof containers, decontaminated, and then safely discharged into the sewer system. Rigid, puncture-resistant, sealable containers are necessary for packaging "sharps," e.g., broken glass, brittle plasticware, needles, and scalpel blades. Wet waste should be packaged with sufficient absorbent materials to contain residual liquids and to minimize leakage. In packaging wet materials for transport, it is prudent to double-bag the waste, sealing each bag independently. Heavy waste such as anatomical specimens, animal bedding, and laboratory specimens need to be placed in rigid containers. Care must be taken that the weight of the waste load does not exceed the burst strength of the container.

The physical properties of the container should be compatible with the treatment process. Waste placed in stainless steel pans, waxed-lined paper bags,

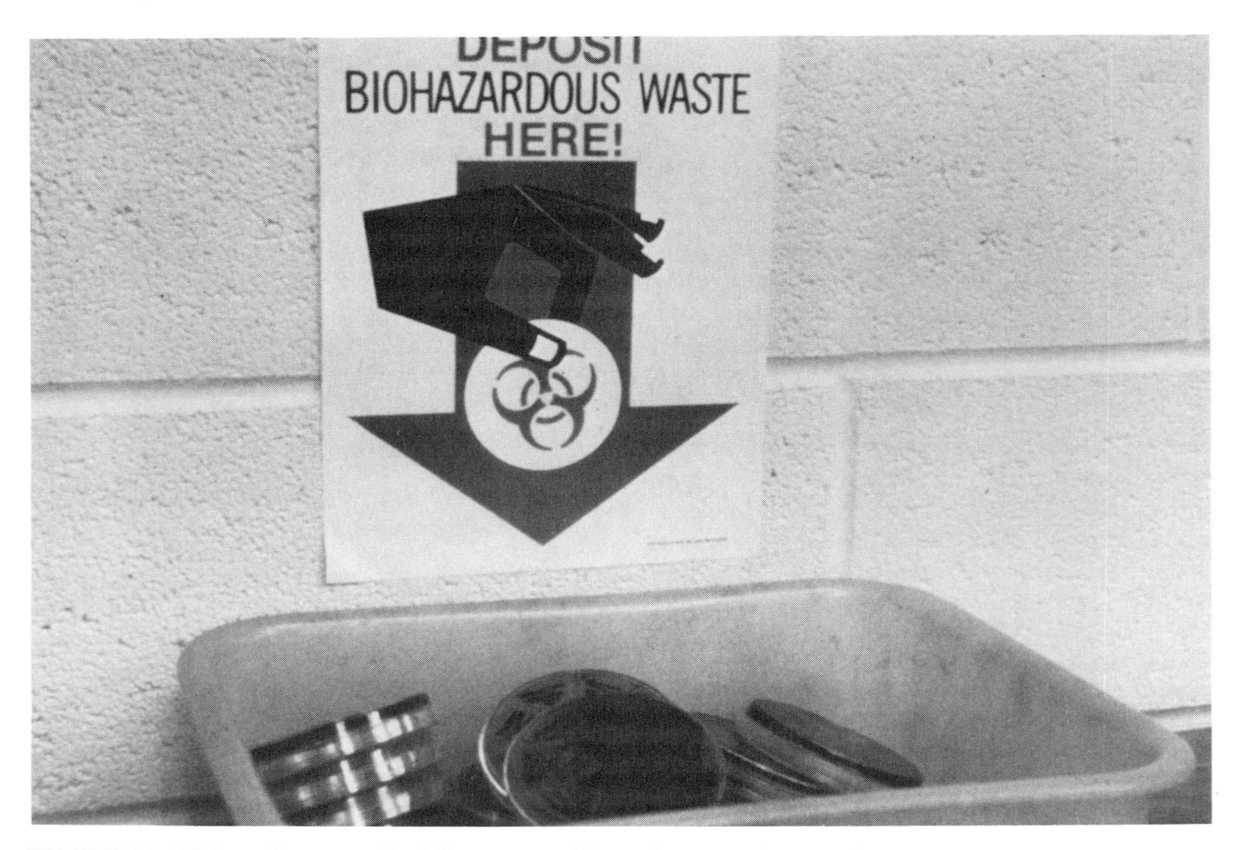

FIGURE 4.1 Biohazardous waste should be segregated from other types of waste prior to its disposal. Courtesy, National Institutes of Health.

tempered glass, and heat-resistant plastics can all be safely processed in an autoclave. Metal containers have been shown to enhance the transfer of heat to the waste load during autoclaving, whereas containers made of plastic retard steam penetration. Processing smaller waste loads and extending the treatment period can compensate for this feature of plastic containers.

Most chemical disinfectants have no appreciable effect on high-strength plastics at room temperature, but may be corrosive to metals. Liquid infectious waste often is stored in plastic carboys designed for chemical disinfection. Metal receptacles can be autoclaved and recycled but are not suitable for incineration. Ideally, waste should be packaged in disposable receptacles that minimize handling of the waste and are suitable for the waste stream treatment method. Cleaning containers that are to be reused is labor intensive and increases the risk of occupational injuries and exposures to biohazards.

3. Personal Protection

The most important precautions for all personnel handling infectious waste are the wearing of protective gloves and frequent handwashing. Gloves and a laboratory coat are recommended for all activities involving manipulations of contaminated items. Gloves and clothing should be changed when soiled or damaged. Thorough handwashing is recommended after working with infectious materials. Scavenging through waste, as well as eating, drinking, and smoking while working with waste, must be prohibited.

The type of laboratory activity will determine if there is a need for additional protective measures. Laboratory activities with a high probability of contamination caused by spills of infectious fluids, or the production of droplets, should be performed on plastic-backed absorbent bench paper. Workers who process infectious waste in an autoclave should wear a rubber apron, sturdy shoes, asbestos-free heat-resistant gloves, and a face shield, to protect against accidents that may occur while loading or unloading the autoclave.

4. Chemical Decontamination

Liquid and gaseous chemicals are used routinely for decontaminating infectious waste. Table 4.1 summarizes use parameters and applications for chemical decontamination of specific types of frequently generated infectious waste from laboratories [140]. Some examples of these applications are as follows:

- Use of an intermediate decontamination step during the storage or transport of waste, e.g., the addition of liquid chlorine bleach, iodophors, or phenolic disinfectants to pipette discard pans at work stations. The concentration of decontaminant for this use should be such that the addition of liquid waste will not interfere with its effectiveness.
- Gaseous decontamination of HEPA filters in biological safety cabinets. This procedure should be carried out prior to removal of the filter for replacement or prior to repairing the cabinet. Decontamination is usually carried out with formaldehyde sublimed by heat from paraformaldehyde flakes in the presence of high humidity. The cabinet must be sealed with plastic sheets and tape prior to initiating decontamination. Human contact with the formaldehyde should be prevented because of the highly irritating, toxic, and possibly carcinogenic properties of the gas (the OSHA limit for permissible exposure is 2 ppm). A detailed description of the method is available [95].
- Decontamination of large items of equipment that are to be removed from the laboratory for repair or discard. Care should be taken to avoid corrosion of sensitive parts if the equipment is to be reused rather than discarded. A disinfectant that has low corrosive properties and has been proven to be effective against the specific microorganism should be used for this purpose.
- Treatment of mixed hazardous waste such as combinations of infectious agents and radioisotopes. After an appropriate assessment of the waste, it may be prudent to use chemically compatible decontaminants to avoid the release of potentially hazardous emissions. See the section on mixed waste (Chapter 4, Section F, Part 1) for a more detailed discussion of such problems.

5. Steam Autoclaving

Steam autoclaving usually is considered to be the method of choice for decontaminating cultures, laboratory glassware, pipettes, syringes, or other small items known to be contaminated with infectious agents. Location of the autoclave within the labora-

TABLE 4.1 Decontaminants and Their Use in Infectious Waste Management

	Ethylene Oxide	Paraformal-dehyde (gas)	Quaternary Ammonium Compounds	Phenolic Compounds	Chlorine Compounds	Iodophor Compounds	Alcohol (ethyl or isopropyl	Formal-dehyde (liquid)	Glutaral-dehyde
USE PARAMETERS									
Concentration of active ingredient	400-800 mg/liter	0.3 g/ft^3	0.1-2%	0.2-3%	0.01-5%	0.47%	70-85%	4-8%	2%
Temperature, °C	35-60	>23							
Relative humidity, %	30-60	>60							
Contact time, min	105-240	60-180	10-30	10-30	10-30	10-30	10-30	10-30	10-600
EFFECTIVE AGAINST[a]									
Vegetative bacteria	+	+	+	+	+	+	+	+	+
Bacterial spores	+	+			±			±	+
Lipo viruses	+	+	+	+	+	+	+	+	+
Hydrophilic viruses	+	+		±	+	±	±	+	+
Tubercle bacilli	+	+		+	+	+		+	+
HIV	+	+	+	+	+	+	+	+	+
HBV	+	+		±	+	±	±	+	+
APPLICATIONS[a]									
Contaminated liquid discard					+			±	
Contaminated glassware	±		+	+	+		+	±	+
Contaminated instruments	±			+				±	+
Equipment total decontamination	±	+							

[a]A + denotes very positive response; ±, a less positive response; and a blank, a negative response or not applicable.

SOURCE: Table adapted from *Laboratory Safety: Principles and Practices*, p. 188 [83].

tory minimizes storage and transport problems. It provides a technically proved treatment method for rendering infectious material safe. Autoclaved waste can be disposed of as general waste.

Certain waste materials are difficult to decontaminate in the autoclave because they insulate and protect the contaminating organisms from heat and steam penetration. Examples include animal carcasses, human body parts, and large volumes of contaminated clothing. The preferred method for decontamination of animal remains and human body parts is incineration. Routine laundering is appropriate for clothing contaminated with all but the most hazardous infectious agents. Autoclaving is not the recommended method for decontaminating very large volumes of waste because the time required for processing is too long, and the chamber size is usually too small. The lack of volume reduction and the failure of the autoclave process to render body parts unrecognizable are also limitations to this process.

Operational considerations based on specific load conditions are very important to ensure adequate decontamination in autoclaves. Most laboratories have gravity displacement autoclaves, which operate at 121°C (15 lbs/in^2 of pressure). Because of the high levels of organic matter normally associated with infectious waste, these types of autoclaves should be operated for a minimum of 60 minutes. Some laboratories may have vacuum-type autoclaves, which operate at 132°C (27 lbs/in^2 of pressure). It is recommended that these autoclaves be operated for a minimum of 10 minutes. The shorter time period for this type of autoclave is due to the higher temperatures and pressures attainable with the vacuum cycle and the more effective penetration of steam.

It may be desirable to add water to a load of waste to be decontaminated in an autoclave to facilitate steam formation and penetration, as well as to avoid the collection of residues on reusable items that may be difficult to remove in subsequent cleaning processes. Caution is essential while adding water to a load, to minimize the potential for aerosolizing infectious agents in the waste. Drain lines from steam autoclaves can be connected to the sanitary sewer except for those installed in maximum containment laboratories (Biosafety Level 4).

When loads contain both reusable and disposable items, the material should be separated to prevent melted plastic from encapsulating items to be reused.

6. Incineration

Incineration is the method of choice for treating large volumes of infectious waste, animal carcasses, and contaminated bedding materials. Because incinerators usually are located some distance from the laboratory, additional precautions for handling and packaging of infectious waste are necessary.

Incinerators require approval and permits from local and state pollution control authorities. Although the initial capital costs and maintenance costs are high, incineration offers many advantages as a method for the treatment of infectious waste [16,49]. Incineration significantly reduces waste volume and produces an unobjectionable end-product, ash. Proper design and operation can provide for energy (heat) recovery, making the operation more economical [25].

Although specific operating standards have not been set for the incineration of infectious or pathological waste, the principles of effective combustion are well understood. Waste and the hot gaseous products of its volatilization should be retained in the combustion chamber(s) for a long enough time and at a high enough temperature to allow for mixing (turbulence) with excess oxygen, so that the combustion reactions can go to completion. A deficiency in any one or more of these critical combustion parameters can result in smoke or odor production, excessive emissions of harmful gaseous by products, and the discharge of incompletely burned waste residue.

Many modern incinerators achieve the proper conditions for complete and effective combustion by providing secondary combustion chambers or zones with burners to ensure that adequate conditions for time, temperature, and mixing are achieved. Primary combustion temperatures of at least 1600°F with good mixing and a gaseous retention time of about 2 seconds should provide for good burnout for the waste described in this chapter. All pathogens and proteinaceous materials are denatured at temperatures well below that just cited [49].

Complete combustion also is dependent on correct operation of the equipment. The operator of the incinerator should be careful to avoid overfeeding with waste materials. Too much raw waste in the primary combustion chamber can overwhelm the combustion zones with more volatile products than the equipment is designed to handle within a fixed gas retention time. The result of overfeeding will be smoke and odors. Overfeeding an incinerator also can result in the bottom ash being moved though the primary chamber too quickly, and consequently being discharged before complete burnout takes place.

Another condition that can result in incomplete burnout of the bottom ash (uncombusted feed material) is the lack of tumbling of the solid waste feed pile in the primary chamber. This is a common condition developing in top-fed incinerators where waste continually is fed directly on top of the existing pile of previously loaded waste materials. In this situation an outer layer of insulating ash can form that retards combustion of the contents in the center of the pile. To achieve complete residue burnout, provisions should be made to agitate or break up the pile periodically. This can be done mechanically with an oxygen pulse or manually with a rake.

In selecting a new incinerator for a facility, it is critical that the actual waste stream to be treated be characterized. Too often in the past the term "pathological waste" has been used to determine the size of an incinerator. True pathological waste consists of animal tissue that is quite wet and has an approximate heat content of 1000 BTU/lb (555.6 kcal/kg). Infectious waste incinerators should burn a wide variety of materials including significant amounts of paper and plastics as well as pathological waste. The effective heat content of the actual waste mix usually will be well above 1000 BTU/lb. Only by knowing the specific composition of the facility's waste stream can a vendor properly size the unit.

In summary, safe, effective incineration can be achieved by (1) proper equipment design; (2) provision for the time, temperature, turbulence, and air required for complete oxidation; and (3) careful feeding of the unit. To assist the laboratory manager in the selection of equipment, a consultant knowledgeable in the field of incineration should be retained to help develop a site-specific procurement specification [49].

7. Validation of Decontamination Methods

Sterility testing or testing for survival of an indicator microorganism is neither applicable nor practical to verify the adequacy of the treatment of infectious waste, since sterility is not an objective of decontamination methods, and indicator microorganisms do not simulate typical waste load composition. Rather, precise reproduction of each of the conditions (operational parameters) prescribed for the different treatment methods should be relied upon to ensure adequate treatment each time waste is processed. This reliance, however, is justifiable only if all of the measuring devices used to monitor the treatment process (e.g., thermometers, pressure gauges, and timing mechanisms) are functioning properly. It is imperative that the accuracy of these measuring devices be certified independently by the user, after the equipment is first installed and before any waste is treated, and again thereafter at regularly scheduled intervals (at least annually). This process should be repeated after maintenance work or repairs are carried out, and if the equipment is relocated.

Each of the different treatment methods for infectious laboratory waste requires a different set of conditions to be effective. Effective autoclaving is dependent upon time, temperature, and steam penetration, whereas effective incineration is dependent upon time, temperature, and turbulence. Chemical decontamination is dependent upon several parameters, including selection of an effective chemical, contact time, concentration, and the presence of organic materials or other interfering substances. Operator controls include such matters as procedures for packaging the waste, placement of the load, feed rate, and—perhaps most important—the keeping of an accurate record of the operational parameters achieved for each load processed. Detailed information regarding the principles of efficacious treatment is available [48,49,52,75,111,140].

Chemical indicators are of limited value in verifying the decontamination of infectious waste. Chemical indicator inks printed on waste packaging materials intended for autoclaving provide a color change that serves only to distinguish treated waste from waste requiring treatment: i.e., failure of an indicator to change to its signal color after the process demonstrates immediately that the equipment has malfunctioned.

F. INFECTIOUS WASTE REQUIRING SPECIAL CONSIDERATION

1. Mixed Waste

Mixed waste consists of materials that exhibit multiple hazardous properties, such as infectivity, radioactivity, and chemical toxicity. Frequently, some of the components are strictly regulated by environmental protection and transportation agencies of the local, state, and federal governments. However, treatment and disposal strategies for mixed waste generally are not addressed by most regulatory agencies. One difficulty, in part due to conflicting regulations, is that mixed waste frequently is not accepted at disposal facilities operating under permits issued by the Environmental Protection Agency (EPA) or licensed by the Nuclear Regulatory Commission (NRC). Waste managers should anticipate that disposal problems will occur. Also, the concern of the public regarding hazardous waste disposal, in general, is likely to influence the development of more restrictive mandates for handling mixed waste. The uncertainty that exists with regard to the treatment and disposal of mixed waste emphasizes the importance of the prudent practice of implementing waste minimization and separation strategies for the types of operations that generate mixed waste.

Waste minimization should be attempted at several levels to address effectively the problem of mixed waste. Researchers should be encouraged to plan experiments and select reagents that minimize the production of mixed waste. Experimental design may be modified such that the wastes are generated separately and in minimal volumes. Microscale techniques are now available for most experimental procedures. When feasible, substitution of less hazardous materials should be considered. Appropriate waste containers should be made available at the work site to ensure convenient and correct segregation and labeling of the waste.

Waste managers commonly have to make a decision about the method to be used for the treatment and disposal of mixed waste. If unusual disposal problems are anticipated by the generator of mixed waste, the waste manager for the institution should be informed prior to generation of the waste. Routinely, the regulated hazardous waste component takes precedence in any treatment and disposal strategy. It is important to recognize, however, that the most serious hazard associated with the waste may not be the regulated component. Placing emphasis on the regulated component therefore may not be the safest approach; in fact, the risks involved in the handling of mixed waste could be exacerbated. A careful assessment of the waste composition is required in order to select the appropriate strategy for treating the mixture. Thermal or chemical inactivation of the infectious component of mixed waste is commonly recommended by waste managers after assessing the potential for toxic emission and chemical incompatibility. Once a strategy for handling the mixed waste is decided upon, it is important that the infectious component be decontaminated along with, or prior to, final treatment and disposal.

Several factors in addition to those associated with the infectious nature of the waste should be considered in selecting an appropriate method of treatment. Some of the more important factors are the type and volume of waste generated; other hazardous properties of the waste (e.g., radioactivity, volatility, flammability, chemical toxicity, and temperature sensitivity [22,91]); the availability of validated treatment methods and recommended safety precautions; the permit and license status of the generator to treat hazardous waste; the quantity and characteristics of the end products after treatment; and the relevant regulations, permits, and packaging requirements for the disposal of the regulated hazardous materials in the mixed waste.

When autoclaving mixed waste, precautions should be taken to avoid the release of volatile radionuclides (e.g., radioiodine) and toxic chemicals (e.g., mercury, solvents, and carcinogens). The potential volatility of the mixture when subjected to the elevated temperatures necessary to achieve thermal biological inactivation is therefore an important factor in the assessment of these types of mixed waste. Mixed waste containing conjugated tritium, technetium-99, carbon-14, and other radionuclides has been autoclaved safely at various institutions, but this method should be approved by the radiation safety officer of the institution. Waste containing flammable or reactive chemicals should not be autoclaved. Care should be taken to avoid contamination of the equipment. Autoclaves used for processing radioactive waste should be labeled with the universal radiation symbol.

Excreta that contain radionuclides from patients undergoing diagnostic nuclear medicine procedures may be discharged into the sanitary sewer if the activity levels in the sewage do not exceed license limitations. Excreta from patients receiving chemotherapeutic drugs may be discharged into the sanitary sewer, if this practice is in compliance with local regulations. Specimens of blood and body fluids obtained from these classes of patients often are submitted for laboratory analysis. These diagnostic specimens can be discharged into the sanitary sewer untreated, or autoclaved and disposed of as general waste.

The incineration of mixed waste is dependent upon the regulatory requirements and permits for the hazardous chemical or radioactive components. Volatile metals (mercury) should not be incinerated but may easily be treated by other methods. Waste management strategies are hindered when the mixture contains volatile metals in addition to other hazardous waste components that dictate incineration as the treatment of choice.

Treatment of all regulated waste and regulated components of mixed waste is restricted to facilities holding the necessary permits. At the time of this writing, none of these facilities accepts waste containing the "dioxin group" compounds (chlorinated phenols and phenoxy acetic acids, chlorinated dibenzo-*p*-dioxins, or chlorinated dibenzofurans) for treatment or disposal. Thus, mixed waste containing any of the dioxin group compounds is difficult to manage. Due to the lack of permitted facilities and the resulting requirement for long-term storage at the producing institution, stringent waste minimization strategies should be implemented to minimize the generation of this type of mixed waste.

Chemical decontamination of mixed waste requires an assessment prior to treatment to avoid potential occupational hazards or difficult disposal problems. For example, decontamination of mixed waste containing radioiodine with sodium hypochlorite could result in the release of the radionuclide.

2. Human Cadavers and Other Anatomical Waste

All anatomical waste requires special handling and packaging due in part to its putrescent properties at ambient temperatures. Cold storage is necessary for nonpreserved anatomical waste to minimize odors and leakage problems. Human body parts, bodies, and cadavers should be disposed of by cremation, burial, or incineration as stipulated in regulations promulgated by the state anatomy board.

Biomedical laboratory and veterinary research operations also generate anatomical waste such as pathological specimens and animal carcasses. Studies of animal and human infections may generate infectious anatomical waste. This waste may be packaged in a durable, preferably opaque, plastic liner that is placed in a sturdy paper fiber box or drum. As a precaution, absorbent material can be added to contain any fluids that might be present. Incineration is the treatment of choice for anatomical waste derived from infected hosts. A facility with a properly designed and operated medical waste incinerator can effectively treat infectious anatomical laboratory and veterinary waste for disposal.

3. Animal Bedding Materials

Soiled animal bedding material also becomes malodorous when stored at ambient temperatures for extended periods of time. Adherence to good sanitation practices can minimize potential problems. Bedding materials obtained from healthy laboratory animals are managed as solid waste and disposed in a sanitary landfill. Institutional or local requirements may apply to the management and disposal of animal waste containing excreta. Decontamination of soiled bedding materials from healthy animals is not indicated, although containment precautions should be instituted to minimize the production of allergenic aerosols.

Contaminated bedding materials and disposable fomites associated with infected animals are also examples of infectious waste. Incineration is the decontamination treatment recommended for combustible, high-density waste. Incinerable small-animal cages are often used in Biosafety Level 3 facilities. This practice avoids the formation of infectious aerosols during the handling and disposal of bedding and related waste. Autoclaves may be available on site or in close proximity to the animal holding areas to decontaminate bedding materials or excreta contained in metal or heat-resistant plastic cages. Decontaminated solid waste then may be managed as general waste.

4. "Sharps"

Needles and other penetrating items such as surgical blades, pipettes, broken glass, and laboratory instrument sampling probes pose a physical hazard to laboratory and support service personnel and, if used to process infectious materials, may transmit infection. All disposable "sharps" should be placed in a prominently labeled, leak- and puncture-resistant container that is consistent with the institution's waste management plan. Sturdy corrugated fiber boxes are used routinely for packaging broken glass and brittle plasticware such as pipettes. The addition of an absorbent material such as crushed corn cob to the package will retain residual fluids and enhance the incineration of waste containing large quantities of plastic and glass. Hypodermic needles and surgical blades often are packaged in disposable buckets made from a high-strength, temperature-resistant plastic. Used disposable needles and syringes should be placed, intact, directly into the waste receptacle without recapping. Other safety design features may include a sturdy handle to transport the waste safely, a restricted opening with a self-activating flap to keep the waste covered while in use, and sealable lids to contain the waste during treatment and disposal. Placing the waste receptacle in a location convenient to the activity generating the waste (e.g., in animal procedure rooms, at patient's bedside, and on the laboratory workbench) is one strategy to ensure proper segregation and packaging of the waste and to minimize the risk of injury to housekeeping personnel.

5

Safety Management

A. ADMINISTRATIVE ORGANIZATION AND RESPONSIBILITIES

1. Introduction

Every institution, irrespective of size, should have a safety program. Such a program should be designed, when appropriate, to ensure compliance with (1) Occupational Safety and Health Administration (OSHA) requirements for health and safety, (2) Nuclear Regulatory Commission (NRC) requirements for safe handling of radioactive isotopes, (3) Environmental Protection Agency (EPA) regulations designed to implement the Resource Conservation and Recovery Act (RCRA), (4) relevant state and local regulations, and (5) requirements of accrediting bodies, such as the Joint Commission on Accreditation of Healthcare Organizations.

Even when a safety program is already in effect, a new laboratory activity may require that the program be modified to address the following issues:

- **the unique hazards introduced by the new activity;**
- **the methods of controlling these hazards;**
- **the new procedures needed (e.g., signs, waste disposal, and personnel monitoring);**
- **the orientation of personnel; and**
- **ways of ensuring that the new procedures are followed.**

Not only should safety programs be a part of an institution's effort, but such an activity should be a central focus of a small office or clinical laboratory as well.

2. The Laboratory Safety Program

a. Goals of a Laboratory Safety Program

The goals of a laboratory safety program should be to protect those working in the laboratory, others who may be exposed to hazards from the laboratory, and the environment. Hazardous materials should be handled and disposed of in such a way that people, other living organisms, and the environment are protected from harm.

b. Responsibility for Laboratory Safety

The ultimate responsibility for safety within an institution lies with its chief executive officer, who, along with all immediate associates (e.g., vice presidents, deans, department heads, laboratory directors, and project directors), should have a continuing, overt commitment to the safety program. This commitment, as well as tangible support, should be obvious to all. A potentially effective safety program that is ignored by top management will fail because it will certainly be ignored by many others. Essential to an effective institutional safety program is a safety coordinator (or safety officer) appropriately trained in relevant safety technology. This individual, besides supplying advice and recommendations, should see that records are kept showing whether the institution's physical facilities and safety rules are internally consistent and compatible with potential risks, as well as with both state and federal laws.

The responsibility for safety in a department or other administrative unit of the institution lies with its chairperson or supervisor. In small institutions, it may be feasible for one person to perform more than one set of duties. For example, a significant fraction of a faculty member's time might be allotted to the duties of the departmental safety coordinator. To be effective, safety coordinators should work closely with administrators and investigators to develop and implement written policies and practices needed for safe laboratory work. Collectively, this group should routinely monitor current operations and practices, see that appropriate audits are maintained, and constantly seek ways to improve the safety program. If laboratory goals dictate operations or substances not suited to the existing facilities, it is the responsibility of the safety coordinator and laboratory supervisor to advise and assist the investigator in developing adequate facilities and appropriate work procedures.

The responsibility for authorization of a specific operation, delineation of appropriate safety procedures, and instruction of those who will carry out the operation lies with the project director.

Taking time to identify potential hazards through a job analysis and to think through its safety aspects is necessary to avoid accidents and illnesses. This practice has proven to be of immense value to industry. Job analysis consists of breaking a job down into its logical steps, analyzing each for its hazard potential, and deciding the safe procedures to use. The process should be designed by a supervisor with input from employees and should be outlined in writing for tasks with potential for injury or other incidents.

Safety awareness should be a part of everyone's habits, and can only be achieved if senior and responsible staff evince a sincere, visible, and continuing interest in the prevention of injuries and occupational illnesses. Laboratory personnel, for their part, must accept responsibility for carrying out their work in a way that protects themselves and their fellow workers.

c. Safety Plans

Because experience has shown that voluntary safety programs are often inadequate, prudent practice requires clearly defined safety rules and monitoring for compliance. These rules should be readily available in writing to all involved in laboratory operations. This goal is often accomplished by preparing a laboratory safety manual containing elements such as outlined below in Section C of this chapter.

Safety plans should be coordinated with institutional and local community emergency services. Discussions with the emergency groups should be held prior to any need for their services, so that they can become familiar with any potential problem areas (e.g., hazardous pathogens, radioisotopes, and chemicals) that may be encountered when they are called for assistance. Telephones or other methods of rapid communication in the event of an emergency should be readily available.

The institution has the responsibility to require that all hazardous materials (e.g., infectious agents, certain chemicals, and radioisotopes) are properly labeled. In addition, persons in the laboratory responsible for handling an emergency should be designated, with telephone numbers posted, so that emergency service personnel and others, such as security guards, know whom to contact at all times of the day or night.

d. Safety Meetings and Safety Committees

In the most effective safety programs, everyone concerned with the laboratory becomes involved. This involvement is usually accomplished by ensuring maximum participation in planning, and by conducting group safety meetings.

In large industrial research laboratories, it is common practice to have monthly meetings of all scientists and technicians reporting to a research supervisor. The chairperson is responsible for developing the agenda of safety topics relevant to the group's activities. Minutes of the meeting are sent to the group members, to the safety coordinator, and to higher management.

In such laboratories, it is customary to have a staff safety committee consisting of the laboratory director and several research supervisors, managers, employees, and the laboratory safety coordinator. The primary purpose of this group is to lead the safety effort, set policy for the group, review accidents and near-accidents, and decide if changes in policies, program, or equipment are needed.

In an academic setting, safety meetings may be held by research groups and by professors and assistants responsible for undergraduate courses. A committee of several professors and the departmental safety coordinator may direct the safety program in the department. Representatives of several departments, including some laboratory technicians, may constitute a committee guiding laboratory safety for the entire institution.

Small laboratories with no formal safety organization should hold periodic safety reviews to discuss actual or potential hazards and how to deal with them, in order to maintain a safety awareness.

e. *Safety Communications*

Safety communications alert people to newly recognized hazards, remind them of basic safety principles, and instill good attitudes toward safety.

Large laboratories often have regular safety newsletters containing useful safety advice and accounts of laboratory accidents along with the lessons to be learned from them.

Safety posters are helpful, but less so than the other kinds of safety communication. Posters should be changed at least every month to catch people's attention.

Reference books on laboratory hazards, occupational health, and good laboratory practices should be readily available. The Material Safety Data Sheets (MSDS) that chemical manufacturers must now supply should also be readily available to all those working in a biomedical laboratory.

The OSHA Hazard Communication Standard (the "right-to-know" rule) [126] requires that every employee be trained to understand the hazards of the substances with which they work, and that current toxicity information be readily available.

f. *Monitoring Safety*

One of the essential elements of a good safety program is the monitoring of the safety performance of a laboratory. Observations of individual safety practices, operability of safety equipment, and compliance with safety rules should be part of the audit. An inspection team, with members selected from several sections of the laboratory, will provide an objective view of the state of safety. Reports of deficiencies and suggestions for correction should go to the people directly concerned. Any malfunctioning facility or equipment should be reported and repaired. Feedback about a particular problem should be brought to the attention of the appropriate supervisor, and, if a problem is widespread, the entire laboratory should be notified.

Essential safety equipment, such as sterilizers and eye wash fountains, should be tested periodically and a record kept of their last inspection. Malfunctioning equipment should be repaired promptly. Personal protective equipment for use in an emergency should be checked periodically, and the qualified users should receive updated training.

B. FACILITIES

1. Introduction

The following discussion presents an overview of the roles of facility design, construction, and maintenance in biosafety. A number of other publications, or references cited therein, address facility design in more detail [19,106,137,145], and others address accreditation [41,72,92]. State and local requirements should also be considered.

For pathogens of veterinary interest, U.S. Department of Agriculture (USDA), Animal and Plant Health Inspection Service (APHIS), and Veterinary Services (VS) personnel should be consulted on facility design. APHIS and VS personnel may elect to inspect the laboratory and animal facilities before a permit is issued to an investigator to begin work with animal pathogens.

The physical facility is a secondary barrier, and it should be designed to ensure a functional laboratory environment that minimizes potential hazards to those working in the immediate area and to others throughout the institution. The design should control traffic, prevent dispersal of aerosols to other areas, and provide for safe movement of hazardous materials and waste. Properly designed physical facilities can provide an environment that is safe, if work procedures are adequate and personal safety devices and primary barriers are properly used. Proper design should, however, allow the laboratory work to be performed in a convenient and cost-effective manner compatible with control of the hazards. Well-designed facilities are also adaptable to changes in equipment and technology or in the kinds of work performed. Resources allocated to the development of the facilities should be commensurate with the risks of the work to be performed. If the facilities are unsatisfactory for the kind of work that is proposed, then either the facilities, the work, or the method of performing the work should be modified.

Before beginning the design of a laboratory in which biohazardous materials will be used, consideration should be given to the known or potential agents to be handled, the procedures to be used, and the quantities of the agents that will be encountered.

Facilities should comply with local, state, and federal building and fire codes, as well as with the requirements of accrediting bodies such as the Joint Commission on Accreditation of Healthcare Organizations, where applicable.

2. Laboratory Design

Biosafety is only one consideration in laboratory design; others include chemical, electrical, fire, and radiation safety. Proper design of a laboratory is a team effort that should include the scientist, engineer, architect, manager, and the safety officer. Too often, an administrator assumes that the architect knows how to design biosafety facilities and that the contractor will construct them properly. In order to minimize costly modifications during and after construction, it is recommended that when new facilities are being built or those extant are being modified, the plans be reviewed by the safety officer and by one or more laboratory scientists to ensure that biosafety requirements will be fulfilled. These individuals should inspect the facility during and at the completion of construction to ensure that plans were followed before the facility is accepted. Final construction (as-built) drawings should be kept available for future reference.

A number of biosafety features should be considered, depending upon the work to be performed in the laboratory. Overall layout of laboratory facilities should consider traffic patterns of personnel and materials; this consideration becomes more important as the biosafety level increases. Facilities should be designed to meet standards described in Appendix A for Biosafety levels 1 through 4.

a. Ventilation

Ventilation is a vital aspect of biosafety. Air handling systems should be such that minimal dust accumulates. Laboratories operating at Biosafety Level 3 or 4 must have directional airflow so that air from these laboratories does not reach other areas. This must be accomplished regardless of the operation of certain equipment, such as biological safety cabinets, that may alter airflow.

There are no uniformly accepted standards of ventilation for Biosafety Level 2 laboratories. Directional airflow is desirable for Biosafety Level 2 laboratories in hospitals. There should be an adequate number of air changes per hour (1) to provide a comfortable work environment, (2) to provide for safe operation of any chemical hoods and biological safety cabinets that are vented directly to the outside or into the exhaust system, and (3) to comply with the appropriate building codes and regulations.

Air from rooms in which biohazardous work is performed should not be recirculated to areas of lesser hazard and is best exhausted to the outside. For example, air from Biosafety Level 3 and 4 areas should not be recirculated. The exhaust air from Level 4 areas should be filtered through HEPA filters and discharged to the outside. In hospitals it is desirable that air from microbiology laboratories and animal rooms not be recirculated, especially to patient care areas. Building codes and regulations in different localities vary and must be followed.

Ventilation exhausts should be remote and not upwind from air intakes, and should not exhaust onto loading docks or patios. If there is a need for filtration or incineration of exhausted air, the integrity of the duct work and proper functioning of the filters or incinerator should be substantiated periodically. It is also important that filters in the ventilation system be changed on a regular schedule.

There should be periodic monitoring of ventilation characteristics to ensure that rooms with special requirements, such as directional airflow or a specified number of air changes, are operating under the proper conditions. Such inspections are particularly important after work has been performed on the ventilating system, including balancing of the system.

b. Electrical

Emergency power needs for laboratories should be defined. Generally, biological safety cabinets in Biosafety Level 2 laboratories will not be on the emergency power system and thus cannot be used if there is a power failure. Circuits providing emergency power should be readily identifiable. Biological safety cabinets that require auxiliary fans on the roof of the building should have an alarm system that will notify the hood operator if the fan fails, or if appropriate negative pressure is not maintained. A plan for action in the event of power failure should be developed.

c. Water

The water supply system should be designed so that back siphonage cannot occur; all faucets that might allow reflux to occur should be equipped with vacuum breakers. If there are both potable and non-potable water sources in the laboratory, each should be clearly labeled. Need for special types of water should be foreseen and, for certain uses, biological contamination should be controlled [88].

Safety showers may be needed for emergencies involving certain biohazards or combined hazards. In most situations when biohazardous material is spilled on one's person, a regular shower in the locker room may be used after clothing is removed. Appropriate handling of the contaminated clothing will depend upon the nature of the hazard. Eye washes should be available in each laboratory handling potentially dangerous material.

d. Sewage

Disposal of waste through the sewage system is frequently an effective way of eliminating material posing a low level of biohazard (see Chapter 4), although, if it is not placed in the drain properly, there is a potential for splatter or aerosolization. It may be important to check with the local sewage plant about its ability to handle waste of certain types or in large quantities, because certain chemicals or biologic products may affect the microbial flora of the treatment facility. Drains should contain sufficient liquid to ensure that the trap is sealed to prevent escape of noxious gases. If a drain is never used, the trap must be filled on a regular schedule or the drain should be sealed.

e. Vacuum

If there is a vacuum system serving multiple areas, care should be taken that there are filters in the system, and that there is an overflow trap containing an appropriate disinfectant to prevent entry of contaminated material into the piping system and pumps. It is often best to use either a stand-alone pump-type vacuum system, or to use a water siphon vacuum system that is attached to a faucet (provided that measures are taken to prevent back-siphonage).

f. Waste Handling

The layout of the physical facilities should facilitate handling of biohazardous waste and should minimize the likelihood of contamination of clean material by such waste [89]. Contaminated material must be segregated from noncontaminated material by physical facilities or appropriate containers.

g. Safety Equipment

Biological safety cabinets, autoclaves, and other biosafety equipment should be properly installed and checked to ensure correct operation. Biological safety cabinets should be certified before use to ensure their operation under appropriate standards (National Sanitation Foundation Standard No. 49 [95]), and should be recertified at least annually (see Chapter 3, Section J). Careful implementation of installation and certification procedures will prevent mistakes, such as leaking filters or inadequate airflow. The laboratory director or safety officer should ensure that employees are properly instructed in the use of safety equipment. The equipment should be recertified if it is moved to another location.

h. Traffic Flow Pattern

The pattern of traffic flow within the facility should be such that the more hazardous areas are remote from other types of operations. Access to a Biosafety Level 3 laboratory through two sequential doors is required. In addition, unnecessary traffic into the laboratory should be discouraged. Doors to laboratories of Biosafety Level 2 or higher should be kept closed when work is in progress in the laboratory.

i. Laundry

Laboratory clothing and towels originating in Biosafety Level 3 and 4 facilities should be routinely decontaminated before being sent to the laundry. Consideration should be given to decontaminating overtly contaminated clothing and towels originating from Biosafety Level 2 laboratories. Clothing and towels from all other laboratories can be sent to the laundry without special treatment.

j. Storage Areas

Storage areas for infectious materials, including stock cultures, actively used infectious materials, and biohazardous waste, should be designed to control access and minimize the likelihood of contamination of personnel or the environment. It is desirable that all hazardous chemicals be stored below eye level.

Freezers, especially liquid nitrogen freezers, present a particular problem because vials or other containers of infectious agents may break and contaminate the liquid nitrogen or a portion of the storage system. Storage in the gas phase of liquid nitrogen freezers is recommended. To decontaminate the freezer, the contents should be removed, the nitrogen allowed to evaporate, and the contaminated areas disinfected and cleaned appropriately.

3. Constructing, Remodeling, and Decommissioning a Laboratory

During construction or remodeling of laboratory areas, there should be careful documentation of the architectural features of the constructed area, and permanent files of the blueprints (as-built drawings) should be maintained. During remodeling, the construction workers should be protected from potential biohazards in work sites. For example, when the dismantling of exhaust ducts from biological safety cabinets in Biosafety Level 3 laboratories is required, the workers should be informed of the potential risk and the ducts should be decontaminated before they are dismantled. If a Biosafety Level 3 laboratory is to be dismantled, or used for other purposes, it must be thoroughly disinfected to ensure that infectious agents from the previous activities no longer constitute a risk to the workers, before any remodeling is carried out. Workers should wear appropriate personal protective devices.

4. Maintenance

The physical plant or engineering group of the organization is generally responsible for maintaining the physical facilities. There should be a defined schedule of maintenance, and if there have been instances of improper or unperformed maintenance in the past, it may be appropriate to insist that records of routine maintenance be made available to the laboratory. It is sometimes helpful for the laboratory also to maintain records of various maintenance activities that are requested and performed. For specific pieces of equipment, records of service calls may aid in justifying replacement. In periodic laboratory inspections, inspectors should specifically review documentation that the maintenance has been properly performed.

Maintenance and physical plant personnel entering a laboratory where work with biohazardous material is being done should either be knowledgeable in proper methods for safely conducting their activities or have proper techniques explained by safety or laboratory supervisory personnel. It is the responsibility of the laboratory supervisor to ensure that the area is decontaminated as needed before any maintenance work or inspections are carried out.

5. Housekeeping

Housekeeping personnel generally do not have a scientific background or a good understanding of various biohazards. It is best that their responsibilities be clearly defined and generally limited to the cleaning of floors, the handling of nonhazardous waste, and other periodic housekeeping activities such as washing walls and windows. It is desirable to develop regular schedules for infrequent housekeeping functions. Working areas such as countertops, reagent shelves, incubators, and refrigerators should be cleaned by laboratory personnel. Spills should be handled by trained personnel, and housekeeping personnel should clean the area only after the infectious hazards have been eliminated. Facilities such as cold rooms, refrigerators, and incubators should have specified laboratory individuals responsible for their regular cleaning.

C. OPERATIONS

1. Introduction

Operations refer to the day-to-day activities of an ongoing safety program. Managing the operational aspects of a safety program requires clear definition of the responsibility and the authority of safety personnel and designation of the chain of command.

This section provides an overview of operations management, but the reader is referred to other sources or to the references cited in them for more detail [61,106,117].

2. Safety Orientation and Continuing Education for Employees

The extent of the orientation and continuing education programs depends upon the size of the organization and the risks to which personnel and visitors could be exposed. Since it is prudent to identify and address all hazards when preparing safety training programs, biosafety should be included along with the chemical safety training required by the new Hazard Communication Standard or the "right-to-know" rule [126] and the radiation safety training requirements of the Nuclear Regulatory Commission. Safety orientation and continuing education should be similar, regardless of whether the work is to be done in research, service, or clinical laboratories, or in academic, private, or government institutions. All employers are responsible for providing appropriate health and safety training for all of their employees. Employees should be motivated to develop a safety awareness and be encouraged to identify unsafe practices or situations in their workplace.

Scientists should undertake projects involving biohazards only if all involved have had the education and training for work at the appropriate biosafety level. The guidelines for microbiological and biomedical research laboratories developed by the Centers for Disease Control and the National Institutes of Health (Appendix A) define the levels of training recommended for each biosafety level. All personnel directly or indirectly involved with the containment and safe handling of known and potentially biohazardous materials should receive instruction and become sufficiently proficient in prudent microbiological practices to allow them to work safely. The curriculum should include instruction in the biology of the microorganisms so that employees can understand the mode(s) of transmission, infectivity, and pathogenicity. It should include also hands-on training in appropriate aseptic technique and decontamination and disposal methods.

Table 5.1 lists a series of topics appropriate for training sessions in biosafety; more extensive outlines of appropriate subjects may be found in Appendix F and reference 149. Documentation by the employee of the completion of orientation and of continuing education is essential.

Training should include information concerning the beneficial aspects of the normal flora of the body, as well as the rarity with which pathogens in the environment cause harm. Such sessions should help to reduce unwarranted fears in the workplace. Information should also be provided on the lack of risk to the general public, since the public perception of the risks of biohazards is far out of proportion to any known problem. Legislative representatives should be educated in a similar fashion, to prevent overreaction to the materials used for research and the waste that institutions have to discard.

Supplementary safety training materials can be obtained from several commercial sources or may be borrowed from government agencies; they need not be developed by the institution. Some examples of sources and programs are listed in Appendix F. Each institution should have a library of reference books and teaching aids, the extent of which will depend upon the financial support and educational expertise available. Consultants can be hired to lecture on specific topics if an expert is not available in-house. To ensure a level of continuity to the program, the project leader or principal investigator may be trained first to serve as an instructor of the technicians on the project. The slide sets and materials in *Fundamentals for Safe Microbiological Research — A Series,* produced in 1983 by the Division of Safety of the National Institutes of Health (see Appendix F), include instruction manuals and advice as to the position of the person in the institution who should address specific subjects. Biosafety officers, sanitarians, microbiologists, infection control practitioners, industrial hygienists, and others may be called upon to give the training. Only those who are knowledgeable about the actual hazards and methods for appropriate controls, and thus capable of answering questions and addressing justifiable and unwarranted fears, should present these programs. Further information on training materials and safety training courses may be obtained from the Division of Safety, National Institutes of Health; Office of Biosafety, Centers for Disease Control; the American Society for Microbiology; the American Biological Safety Association; and the Biohazards Section of the American Industrial Hygiene Association. Physicians' laboratories may contact their state health laboratories for further information.

TABLE 5.1 Suggested Topics for Biohazard Safety Training

1. Aseptic technique and procedures
2. Personal hygiene
3. Laboratory practices (primary containment barriers) for appropriate biosafety levels
4. Personal protective equipment
5. Assessment criteria for facilities at various biosafety levels
6. Decontamination, disinfection, and sterilization
7. Signs
8. Biohazardous waste handling, packaging, and disposal
9. Packaging, transporting, and shipping biohazardous materials
10. Effective use of a biological safety cabinet
11. Safe use of an autoclave
12. Safe use of a centrifuge
13. Monitoring and auditing
14. Reporting incidents and accidents
15. "Right-to-know" hazard communication
16. "Universal precautions" for handling human blood and body fluids

NOTE: See Appendix F for sources of audiovisual materials.

Graduate degrees in biosafety are offered through the Biohazard Science Training Program, School of Public Health, University of North Carolina. Other schools of public health offer advanced degrees in environmental health or epidemiology, which can include elective or required courses in biosafety. Short courses or workshops in biosafety or related subjects are sponsored by national associations such as the American Society for Microbiology, Association of Practitioners in Infection Control, and the American Chemical Society, or as continuing education programs through academic institutions. Training programs sponsored from 1980 through 1983 by the National Institutes of Health have continued to be offered annually by the host institutions, Harvard University, and The Johns Hopkins University, on a fee basis, to meet the demand for such training (see Appendix F).

3. Evaluation of Laboratory-Associated Hazards

The dangers to personnel and the environment from biohazardous laboratory activities should be assessed in a systematic fashion. A number of factors should be considered, but the two most important factors are (1) the agents and (2) the potential consequences of infection. The characteristics of the microbiological agents being used are particularly important; i.e., their virulence, pathogenicity, communicability, and route of spread are properties affecting the potential danger for laboratory workers and the environment. The types of procedures used with the microorganisms and the quantities handled will also affect the degree of hazard. Agents causing infections that are mild, easily treated, or readily prevented with a vaccine pose much less danger than agents that cause severe or fatal disease and cannot be prevented or treated effectively. In addition, the skills and knowledge of the employees should be considered. The individuals working in a laboratory should be trained adequately to understand the hazards of their work, to become proficient in the procedures that should be followed to minimize personal danger, and to be aware of the possibility that they might expose others to the organisms.

Assessment of hazard requires good judgment in the application of general principles to the specific laboratory situation to reach a rational decision. It includes also an evaluation of the types of facilities, laboratory practices, personal protection devices, and equipment that will be needed to perform the laboratory work safely.

4. Policy and Procedure Manuals

It is essential that biosafety policies and procedures be clearly spelled out in a manual, including the information the laboratory worker should know for day-to-day activities as well as for handling emergencies. The laboratory biosafety manual should include the following subjects:

- **policy and goals;**
- **safety organization;**
- **medical program;**
- **procedures for general laboratory operations, including: labeling and handling of speci-**

mens; methods to minimize hazards of aerosols and droplets; proper uses of needles, syringes, and "sharps"; appropriate discard of working materials; sterilization and disinfection; cleanup of spills; use and maintenance of biological safety cabinets; control of insects and other pests; work with animals; and waste disposal;

- **safety equipment location and proper use; and**
- **emergencies.**

The safety manual should be readily available to all employees. Each new employee should be required to review the manual and to document that this has been done. The safety manual should be reviewed annually by the laboratory director and by the safety officer to ensure that it is accurate and current for the laboratory in which it is being used.

In addition to a safety manual, many laboratories have a manual of general procedures (standard operating procedures, or SOPs). This manual should include the special safety precautions required for particularly hazardous steps of the various procedures described therein. Safety aspects of each procedure should be reviewed during the periodic review of the SOPs.

Work that involves animals or the use of animal facilities may require additional safety procedures, which should be clearly defined in procedure manuals. These are described in more detail in other publications [44,105]. (See Appendix G, Accreditation.)

Smaller laboratories, such as physicians' office laboratories or laboratories that handle minimally biohazardous materials, may address biosafety as a portion of the general laboratory procedure manual. For many small laboratories, collection and handling of specimens of blood and body fluids represent the major hazard.

5. Accident Reports and Investigations

Each organization should have a defined system for reporting laboratory injuries and accidents, as well as for investigating them. These events should be documented and reported to the appropriate supervisory personnel and to the employee health service. For those organizations subject to the regulations promulgated by the Occupational Safety and Health Administration (OSHA), there are specific requirements for reporting injuries in the workplace. There may be requirements for similar reporting by state and local governments.

Most large organizations will have special forms for recording accidents. However, in smaller organizations an expository description will suffice for most accidents. Reports should be filled out for all laboratory accidents. These should include a description of the accident and any factors contributing to the accident. In addition, any first aid or other health care given to the employee should be included. Responsibility for completing these forms should be clearly defined. It might be assigned to the laboratory supervisor, employee health personnel, or safety personnel. Accidents should be periodically reviewed by the safety committee, by the employee health unit, or by other appropriate personnel, and the individual reports or a summary should be sent to the director of the organization.

The information collected from accident reports can be used to investigate specific accidents and can be collected and analyzed to assess trends in various problems. For example, is the frequency and/or severity of needle sticks increasing or decreasing? Have changes in procedures for handling needles affected the incidence or severity of such accidents?

6. Recordkeeping

A number of records of the biosafety program are required by OSHA, and perhaps by the local and state government. In general, records should be kept for a minimum of 5 years, although some, such as medical records, should be held for 30 years. Modern data processing can simplify many aspects of biosafety recordkeeping.

7. Auditing

Auditing (quality assurance) is essential if the safety program is to function properly. If there is no effort to ensure that appropriate procedures and policies have been carried out conscientiously, the program may exist only on paper. It is helpful to have a periodic external review by a safety officer, a person who works in another laboratory, or an inspector from an accrediting organization. Examples of the latter for health care laboratories are the Joint Commission for Accreditation of Healthcare Organizations, the College of American Pathologists, and state health departments. These agencies have lists of safety and other laboratory requirements and will check to see if these are fulfilled. Facilities, equipment, and employee knowledge of safety and compliance are some of the items reviewed. Internal auditing reviews of the laboratory, using a checklist obtained from such an accrediting group, also may be helpful.

It is important to conduct periodic audits of key physical aspects of the biosafety system. For example, during these reviews the annual certification of biosafety cabinets and periodic testing of the ventilation and alarm systems can be confirmed. This practice is particularly important after modifications in the facilities have been made as a result of periodic maintenance or renovation. It is not unusual to find that supposedly negative pressure rooms are under positive pressure or have inappropriate numbers of air changes per hour. Reports of the results of the audits should be submitted to management for review and action.

8. Registry of Agents

In any large organization, a central registry should be maintained of the identity and location of the various infectious agents being handled throughout the facility, particularly for those agents requiring Biosafety Level 3 or 4 operations. A central registry is essential for dealing with emergencies. For example, a fire in a laboratory might cause excessive damage if the firemen were unwilling to enter an area marked only with a conventional biohazardous or other type of warning sign if no one was readily available who could explain the nature of the potential hazard within that laboratory. It may be helpful to list, on the hazard warning signs on the laboratory doors, the agents, the common names of the diseases caused by them, and the names and telephone numbers of persons to be contacted in the event of an emergency.

9. Waste Management

It is the responsibility of the laboratory director to see that waste is properly handled and, when necessary, that it has been made noninfectious before being discharged into the environment. The principles of infectious waste management are described in Chapter 4.

10. Signs

Appropriate signs should be used to identify hazardous laboratory areas as outlined in Appendix A (see section on Registry of Agents above). These biohazard signs should be posted at the entrances to areas if there are special conditions for entry. When there are multiple potential hazards, multiple signs are required. There should be a uniform system of signs within an institution, and it should conform with nationally recognized symbols.

D. MEDICAL PROGRAM

1. General Principles

The extent of the medical program for employees with potential exposure to infectious agents should be based upon the specific risks and hazards of the laboratory activities, as well as on the overall medical program of the organization of which the laboratory is a part. Except for investigations of accidental exposures or inadvertent infections, the main objective of the medical program should be to prevent disease. It is beyond the scope of this book to discuss the role of the medical program in promoting the general health of employees (e.g., screening for diabetes, high blood pressure, and heart disease).

Designing the program requires clear definition of goals and an appropriate plan to reach these goals [57,58,135]. Components of a medical program might include: preplacement examination (PPE), periodic monitoring evaluation (PME), treatment and documentation of accidental exposure, epidemiologic study of exposure-related illness, tracking of prolonged or unusual illness, immunization, and postemployment evaluation (PEE) [62]. The program might also be involved in making recommendations to reduce exposure to biohazards (e.g., use of biological safety cabinets, use of personal protective devices, or changes in work practices).

2. Conditions Increasing Employee Risk of Adverse Health Outcome

The risk of adverse health outcome as a result of exposure to infectious agents may be increased in individuals whose immunological or other defense mechanisms have been impaired by such conditions as medication, allergy, or pregnancy. When such conditions are identified in employees potentially at risk, it is important to review the nature of their work to ascertain whether or not some change or accommodation can be made to lower or eliminate the chances for exposure. If such actions are not possible, it may be necessary to transfer the employees to other jobs. In the case of a prospective candidate, such findings may be the basis for denying employment. These decisions must be individualized with input from the employee, management, the institution's occupational medicine service, and — if appropriate — the employee's private physician.

a. Deficiencies of Host Defenses

The considerable literature on the infection of persons working in microbiological laboratories has been described in Chapter 2, Chapter 3 (Section B), and Appendix A. Because of this occupational risk it is reasonable to conclude that, in addition to the need for engineering controls, biosafety equipment, appropriate work practices, and personal protective devices, laboratory workers should have unimpaired host defenses. A detailed discussion of abnormalities of host defenses is beyond the scope of this document, but some major factors are briefly considered below.

Cutaneous defenses are altered by diseases such as chronic dermatitis, eczema, and psoriasis. Persons with these conditions may be more susceptible to skin infections. A disrupted skin surface may allow entry of such agents as hepatitis B virus and human immunodeficiency virus, which are believed not to be able to penetrate healthy, intact skin.

Antimicrobial therapy may interfere with protection afforded by the normal microbial flora of the mucosal surfaces of the body. Antibiotic-related suppression of the normal flora, or bowel pathology that disrupts the mucosal surface, may interfere with the protective properties of the healthy, intact gastrointestinal tract.

Abnormalities of the immune system may interfere with antibody-mediated defenses, T-cell-mediated defenses, phagocytosis, or complement-mediated defenses. The types of infections likely to occur will vary with the nature of the alteration of the host's defenses. For example, persons who have asplenia, complement defects, antibody defects, or decreased numbers of polymorphonuclear leucocytes are more likely to be subject to serious infection caused by encapsulated bacteria (e.g., pneumococcus or haemophilus), while those with T-cell defects are at a greater risk of developing active tuberculosis, histoplasmosis, listeriosis, or cytomegalovirus pneumonia. Treatment with corticosteroids can interfere with T-cell, B-cell, and phagocyte functions.

b. Reproductive Hazards

The laboratory environment may contain various biological, chemical, or physical hazards that can adversely affect the outcome of pregnancy, but this discussion will be limited to the biological hazards. Although sexually acquired infections with some agents (e.g., chlamydia and gonococcus) can result in infertility in either sex, it is unlikely that accidental infection with these agents in the laboratory will cause this problem. Exposure to mutagenic agents would be of concern to fertile employees of either sex, and such exposure should thus be minimized.

Of special concern is the potential for infection of the fetus, in utero or during delivery, resulting from a work-related infection acquired by a pregnant employee. Diagnostic microbiologists, serologists, and chemistry laboratory workers who have no direct contact with patients with infectious diseases may be exposed unknowingly to a variety of infectious agents in the specimens that they process. The infectious agents that potentially may be acquired in this manner and that are known to cause congenital or neonatal infections include rubella virus, hepatitis B virus, cytomegalovirus, human immunodeficiency virus, enteroviruses, herpes simplex virus, varicella virus, *Treponema pallidum* (the agent of syphilis), and toxoplasma [112]. The concern for the pregnant employee may be increased if she is handling viruses requiring Biosafety Level 3 or 4, for which the effects of maternal infection on the fetus are unknown.

When considering job placement or work scope for a pregnant laboratory employee, or an employee who is attempting to become pregnant, several factors are important: the agent, and what is known about the risk of infection and its consequences; the means available to prevent exposure; and the possibility of other assignments [144]. The employee should be completely informed of the potential risks, and should be involved in the decision-making process along with the employer and the employee's physician. In making a decision, consideration must be given to both the health of the fetus and the needs of the employee.

c. Allergies

Workers in biomedical laboratories may develop allergies to aerosolized proteins (e.g., fermentation products and enzymes), animal dander, urine proteins [5,14], or arthropod materials (see Chapter 3, Section C). In addition, allergies to substances (such as egg proteins) found in vaccines may preclude immunization. Specific allergies may occasionally be grounds for denying employment or job transfer or may necessitate the institution of stringent measures to prevent exposure.

3. Program Design

After the medical program's goals have been defined and the potential hazards recognized, the content of the program and its participants can be determined. The program should be in compliance with federal, state, and local government regulations for the protection of the health and welfare of the employee (see Section F of this chapter), and should be in accord with the general policies of management.

4. Preplacement Examination (PPE)

The primary goal of a preplacement examination (PPE) for a prospective employee is to reveal any medical condition that might put the worker or coworkers at increased risk because of certain job exposures or activities. Recommendations for job restrictions or denial of employment should be carefully considered and should avoid exclusions based on inaccurate presumptions of risks or future disabilities. Other goals of the PPE are to obtain a baseline for future comparison and to ensure that the employee has received appropriate immunizations.

At the time of hiring, it is prudent to give all prospective employees a questionnaire that elicits an occupational and medical history. Workplaces not affiliated with an occupational medical service could have this done through the employees' own physicians.

Personnel who may require a more complete evaluation (e.g., physical examination and laboratory testing) include the following:

- individuals working with the more hazardous agents, particularly those requiring Biosafety Level 3 or 4 (see Appendix A and reference 134);
- individuals working with oncogenic and teratogenic microbial agents;
- animal caretakers, maintenance personnel, custodial and housekeeping staffs, or others who work in areas where potential exposure to infectious materials is high;
- personnel working in overcrowded areas or in areas where containment systems are deficient or lacking; and
- individuals whose questionnaires show medical conditions that would increase their risk when handling certain agents.

a. Medical History

The medical history should elicit information about the state of general health and host defenses, the presence of allergic conditions, past immunizations, and prior significant infectious diseases. It may be useful also to inquire about the presence of specific symptoms (e.g., persistent nasal congestion, headache, cough, or gastrointestinal complaints), which may be useful in the future for evaluating medical complaints of the employee and for interpreting the significance of nonspecific early manifestations of illness.

b. Occupational Health History

The occupational health history is the cornerstone of the occupational health examination [57,58]. The initial occupational history should include a profile of past employment; a listing of symptoms, illnesses, or injuries related to past hazardous occupational exposures; specific questions relevant to the requirements and potential hazards of the proposed work assignment; and any significant factors concerning community and home contacts and activities.

c. Physical Examination

If a general physical examination is performed, particular attention should be paid to evidence suggesting altered host defenses (e.g., lymphadenopathy, hepatospenomegaly, dermatitis, or allergy).

d. *Laboratory and Other Testing*

Biological monitoring and medical screening can be key elements in a medical program. Biological monitoring provides evidence of infection with specific infectious agents, e.g., the purified protein derivative (PPD) skin test for tuberculosis and serological tests for antibody to hepatitis B virus. Medical screening detects evidence of abnormal structure or function of body organs and systems, e.g., chest radiographs, total and differential white blood cell counts, and liver function tests. The purpose of biological monitoring and medical screening is to identify the occurrence of inapparent infections and to permit recognition of the early signs and symptoms of disease when the effects of the infection may be reversible or more easily treated [62]. Selection of the laboratory testing procedures should be based upon the potential hazards, whenever possible.

Skin testing of clinical microbiologists with purified protein derivative for tuberculosis can be useful. If the PPD skin test has recently converted from negative to positive, a chest radiograph is indicated to detect active disease. Periodic chest radiographs have not been of value in monitoring asymptomatic employees with long-standing positive PPD skin tests [103,112]. It is rarely necessary to do a chest radiograph in an otherwise healthy person who has no pulmonary symptoms and whose physical examination and PPD skin test are negative.

Preplacement patch testing for allergies has not proved to be a generally effective means of predicting a predisposition to the development of allergies in animal handlers.

Pulmonary function testing should be done only in special circumstances, such as assessing the fitness of an individual to wear a respirator.

In work situations where the competency of host defenses is important, one might consider performing other tests, such as a complete blood count with differential white cell count, which might indicate the presence of conditions compromising host defenses.

e. *Serum Bank*

A decision to establish a serum bank should be considered carefully. A serum bank may be helpful in finding evidence for presumptive infection with certain infectious agents and in establishing the approximate time of its occurrence. The potential usefulness of a serum bank will depend upon the agents to which the employees may be exposed, the likelihood that apparent or inapparent infection will cause a change in the level of humoral antibodies, and the availability of diagnostic tests for the agents being handled.

All serum specimens should be stored frozen. Comparison of the levels of antibody or antigen in the initial specimen with those in subsequent specimens allows detection of changes. Testing is usually not performed until there are two or more specimens that can be assayed simultaneously to ensure comparability of results. The serum bank can be invaluable when there are questions of job-acquired infection.

5. Immunizations

The recommended childhood immunizations for diseases such as tetanus, diphtheria, pertussis, poliomyelitis, measles, mumps, and rubella are of particular benefit to clinical laboratory workers and to those working with the etiologic agents causing these diseases or with specimens that may contain them (see Chapter 3, Section B). Maintaining tetanus immunization simplifies care of penetrating injuries. Persons with patient contact should have immunity to rubella to protect themselves, as well as to prevent transmission of this infection to others who may be pregnant.

Specific immunizations may be desirable for employees at high risk of exposure to certain agents. All laboratory workers who are exposed to human blood or body fluids should be immunized against hepatitis B virus. Rabies immunization is appropriate for persons working with rabies-infected specimens and animals (see Table 5.2). For more specific and detailed recommendations, the reader should consult the latest edition of the American College of Physicians' *Guide for Adult Immunizations* [2].

6. Periodic Monitoring Examination (PME)

The goals of the periodic monitoring examination (PME) include the following:

- detection of the effects of possible exposure to biohazards (e.g., symptoms of clinical infection; allergic symptoms; or complaints related to the pharmacological effects of end-products, by-products, medium components, or inactivated biological agents);

- detection of changes in the health status of the employee since the last evaluation that may indicate the need for a change in work procedures or job placement; and
- detection of patterns of disease in the work force that may indicate work-related problems and suggest the need to evaluate the effectiveness of control measures.

The frequency of the PME varies depending on the employment situation, but usually is performed annually. The PME might include an update of the occupational and medical histories, biological monitoring tests to detect subclinical infection, medical screening tests, and targeted physical examination.

7. Postemployment Evaluation (PEE)

A postemployment evaluation (PEE) may be desirable immediately before an employee leaves the laboratory of the employer. This evaluation should resemble the PPE and include, at the least, an interval occupational medical history. The extent of the evaluation can be modified if the person has undergone either a PPE or PME within the previous six months. If a serum bank has been established, it may be appropriate to obtain a final serum sample.

8. Agent-Specific Surveillance

The implementation of a surveillance system may be useful for the early detection of symptoms related to a specific etiologic agent or to the effects of exposure to noninfectious products in the workplace. For example, an ongoing surveillance program for all workers who may become exposed to rickettsiae in the laboratory, including maintenance and other support service personnel, could result in early treatment of the disease and amelioration of its severity (see Appendix C in reference 105). Such a surveillance system should include the availability of an experienced medical officer, education of at-risk personnel about the potential hazards of the infectious agents and the advantages of early treatment, a reporting system for recording all recognized exposures and accidents, and the requirement for prompt reporting of all febrile illness.

9. Accident Reporting

Accidents, such as needle sticks or spills, should be reported immediately to the laboratory supervisor. The supervisor should refer the employee to the medical staff or consulting physician to determine whether testing, treatment, or follow-up is needed. Accidents should also be reported to the biosafety staff, so that appropriate measures can be instituted to avoid a similar accident in the future. Protocols for preventing anticipated accidents, such as needle sticks and spills, should be prepared, and employees should be trained to deal with these situations accordingly. Laboratory personnel can be trained as "first responders" for dealing with various medical emergencies [50].

Accident reporting and recording should be in compliance with legal requirements (e.g., OSHA regulations).

10. Recordkeeping and Result Notification

Workers should be informed of the results of their occupational medical evaluations. Confidentiality of medical records and test results should be maintained, and information released to others only with the authorization of the involved employee. Authorization for the release of medical information that is pertinent to their fitness to perform their job duties may be obtained from the workers at the time of their PPE. The personnel officer, or the biosafety officer, need only receive information about those medical conditions, restrictions, or accommodations that relate specifically to the individual's fitness for work. (See the Americal Occupational Medical Association *Code of the Ethical Conduct for Physicians Providing Occupational Medical Services* [7].)

It is advisable to review periodically the data collected in the medical surveillance program. In larger laboratories, for example, the prevalence rates of symptoms or abnormalities in different job or exposure groupings can be examined, and the results of medical monitoring data can be related to the laboratory environmental sampling data.

11. Resources

The medical program should be designed and supervised by a qualified individual, such as a physician or nurse practitioner. Depending upon the size

TABLE 5.2 Recommendations for Immunoprophylaxis of Personnel at Risk

Disease	*Description of Product*	*Recommended For Use In*	*Source of Product*
Anthrax	Inactivated vaccine	Personnel working regularly with cultures, diagnostic materials, or infected animals	USAMRIID[a]
Botulism	Pentavalent toxoid (A,B,C,D,E) IND[b]	Personnel working regularly with cultures or toxin	CDC[c]
Cholera	Inactivated vaccine	Personnel working regularly with large volumes or high concentrations of infectious materials	Commercially available
Diphtheria-Tetanus	Combined toxoid (adult)	All laboratory and animal care personnel irrespective of agents handled	Commercially available
Eastern Equine Encephalomyelitis (EEE)	Inactivated vaccine	Personnel who work directly and regularly with EEE in the laboratory	USAMRIID[a]
Hepatitis A	Immune serum globulin (ISG [Human])	Animal care personnel working directly with chimpanzees naturally or experimentally infected with hepatitis A virus	Commercially available
Hepatitis B	Serum-derived or recombinant vaccine	Personnel working regularly with blood and blood components	Commercially available
Influenza	Inactivated vaccine	(Vaccines prepared from strains isolated earlier may be of little value in personnel working with recent isolates from humans or animals)	Commercially available
Japanese Encephalitis	Inactivated vaccine	Personnel who work directly and regularly with JE virus in the laboratory	CDC[c]

Disease	*Description of Product*	*Recommended For Use In*	*Source of Product*
Measles	Live attenuated virus vaccine	Measles-susceptible personnel working with the agent or potentially infectious clinical materials	Commercially available
Meningococcal Meningitis	Purified tetravalent polysaccharide vaccine	Personnel working regularly with large volumes or high concentrations of infectious materials (does not protect against infection with group B meningococcus)	Commercially available
Plague	Inactivated vaccine	Personnel working regularly with cultures of *Yersinia pestis* or infected rodents or fleas	Commercially available
Poliomyelitis	Inactivated (IPV) and live attenuated (OPV) vaccines	Polio-susceptible personnel working with the virus or entering laboratories or animal rooms where the virus is in use	Commercially available
Pox viruses (Vaccinia, Cowpox, or Monkey Pox viruses)	Live (lyophilized) vaccinia virus	Personnel working with orthopox viruses transmissible to humans or with animals infected with these agents, and personnel entering areas where these viruses are in use	CDC[c]
Q Fever	Inactivated (Phase II) vaccine	Personnel who have no demonstrable sensitivity to Q fever antigen and who are at high risk of exposure to infectious materials or animals	USAMRIID[a]
Rabies	Human diploid cell line inactivated vaccine	Personnel working with *all strains* of rabies virus or with infected animals, and personnel entering areas where these activities are conducted	Commercially available

Continued

TABLE 5.2 *Continued*

Disease	*Description of Product*	*Recommended For Use In*	*Source of Product*
Rubella	Live attenuated virus vaccine	Rubella-susceptible personnel, especially women, working with "wild" strains or in areas where these viruses are in use	Commercially available
Tuberculosis	Live attenuated (BCG) bacterial vaccine	(BCG vaccine ordinarily is not used in laboratory personnel in the U.S.)	Commercially available
Tularemia	Live attenuated bacterial vaccine (IND)[b]	Personnel working regularly with cultures or infected animals, and personnel entering areas where the agent or infected animals are in use	CDC[c]
Typhoid	Inactivated vaccine	Personnel who have no demonstrated sensitivity to the vaccine and who work regularly with cultures	Commercially available
Venezuelan Equine and Related Encephalitides (VEE)	Live attenuated (TC83) viral vaccine	Personnel working with VEE and the Equine Cabassou, Everglades, Mucambo, and Tonate viruses, or who enter areas where these viruses are in use	USAMRIID[a]
Western Equine Encephalo-myelitis (WEE)	Inactivated vaccine	Personnel who work directly and regularly with WEE virus in the laboratory	USAMRIID[a]
Yellow Fever	Live attenuated (17D) virus vaccine	Personnel working with virulent and avirulent strains of yellow fever virus	Commercially available

[a] U.S. Army Medical Research Institute for Infectious Diseases, Fort Detrick, MD 21701, telephone: (301) 663-2405.
[b] Investigational new drug.
[c] Clinical Medicine Branch, Division of Host Factors, Center for Infectious Disease, Centers for Disease Control, Atlanta, GA 30333, telephone: (404) 639-3356.

SOURCE: Adapted from recommendations of the PHS Immunization Practices Advisory Committee and *Biosafety in Microbiological and Biomedical Laboratories* [105].

of the laboratory (or company), the director of the medical program may be full-time and on site, such as at a facility-based employee health service, or part-time and off site, where employees are evaluated at a different location. Whatever the arrangement, there should be an open line of communication between the medical program director and the person(s) responsible for biosafety. The medical program director must be familiar with the nature of the employees' work and its potential hazards in order to design an effective medical program.

E. EMERGENCIES

1. Preparation and General Procedures

a. Preparation

It is the responsibility of every laboratory organization to establish a specific emergency plan for its facilities. This plan should cover both the laboratory building and the individual laboratories. For the building, the plan should describe evacuation routes and shelter areas, facilities for medical treatment, and procedures for reporting accidents and emergencies. It should be reinforced by drills and simulated emergencies. Plans should include liaison with local emergency groups as well as with community officials. To be prepared, these groups should be informed of plans in advance of any call for assistance.

"Community right-to-know" regulations [127] require the development and coordination of emergency plans with local community response groups, as well as the listing of hazardous chemicals and their location. Many types of facilities are included in these requirements; however, there are exemptions for qualifying laboratories. Legal guidance should be obtained.

For small-scale accidents in the laboratory, a good "rule of thumb" to remember is to leave the area, call for help, and then secure the area. Preplanning of work is the best way to avoid accidents and should include thinking through "what if" an accident should occur unexpectedly. In handling mixed hazards (e.g., a substance that may be infectious and radioactive, or infectious and chemically toxic, or present all three hazards) it is usually best to respond with procedures for the greater hazard first, and then follow through with those for the lesser hazards, to ensure that all appropriate steps have been taken. General emergency procedures are discussed below, followed by guidelines for dealing with specific types of accidents.

b. General Emergency Procedures

The following emergency procedures are recommended in the event of fires, spills, explosions, or other laboratory accidents. These procedures are intended to limit injuries and minimize damage if an accident should occur.

- **Render assistance to persons involved and remove them from exposure to further injury if necessary; do not move an injured person not in danger of further harm.**
- **Warn personnel in adjacent areas of any potential hazards to their safety.**
- **Render immediate first aid (e.g., beginning resuscitation if breathing has stopped or washing under a safety shower). Appendix 4 of reference 83 contains an Emergency First Aid Guide.**
- **In case of fire, call the fire department. Follow local rules for dealing with a small fire, e.g., if there are portable extinguishers available and the institution encourages their use, extinguish the fire. On the other hand, some institutions require all fires to be reported immediately, thereby summoning trained assistance.**
- **In a medical emergency, summon medical help immediately. Laboratories without a medical staff should have personnel trained in first aid available during working hours.**

2. Evacuation Procedures

The following evacuation procedures should be established and communicated to all personnel.

a. Emergency Alarm System

There should be a system to alert personnel of an emergency that may require evacuation. Laboratory personnel should be familiar with the location and operation of alarm equipment.

Isolated areas (e.g., cold, warm, or sterile rooms) should be equipped with alarm or telephone systems that can be used to alert outsiders to the presence of a worker trapped inside, or to warn workers inside of the existence of an emergency that requires evacuation.

b. Evacuation Routes

Evacuation routes should be established. An outside assembly area for evacuated personnel should be designated, with plans for taking roll call to ensure that all personnel are accounted for.

c. Shutdown Procedures

Brief guidelines for shutting down operations during an emergency or evacuation should be available in writing. Biohazardous agents should be secured in cabinets to minimize the danger of spillage.

d. Start-Up Procedures

Written procedures to ensure that personnel do not return to the laboratory until the emergency is ended, and start-up procedures that may be required for some operations, should be displayed and reviewed regularly.

e. Drills

All aspects of the emergency procedure should be tested regularly (e.g., every 6 to 12 months).

f. Power failure

Loss of power can result in failure of a containment system, or loss of lighting, ventilation, refrigeration, or other essentials for safety. Procedures to handle this type of emergency should, therefore, be included in planning.

3. Fires

Fires within a laboratory using biohazardous materials will require an immediate response to minimize personal exposure and to limit the potential spread of biological contamination. Fires may create toxic smoke as well as aerosols that may contain infectious materials.

Small fires that may occur in a laboratory usually are extinguished by the immediate use of a portable fire extinguisher. A Halon extinguisher is preferred for use because it extinguishes the fire quickly without leaving any chemical residue to contaminate the work area. Water from the sprinklers of the building fire protection system should extinguish the fire. In case of doubt about containing a fire, no time should be wasted in deciding to call the institutional or community brigade. Fire fighters responding to the fire scene should wear self-contained breathing apparatus to protect themselves from toxic combustion byproducts and aerosols generated by the burning of infectious materials.

4. Spills and Releases

Experience has shown that the accidental spill and release of hazardous substances is a common enough occurrence to require procedures that will minimize exposure of personnel and contamination of property. Such procedures may range from having available a sponge mop and bucket to having an emergency spill-response team, complete with protective apparel, safety equipment, and materials to contain and clean up the spill. In any event, there should be supplies on hand to deal with the spill consistent with the hazard and quantities of the spilled substance.

a. Infectious Agents

Some biological research materials, when dropped, spilled, or set on fire, can release hazardous agents that can contaminate the area and lead to infection of laboratory workers. Prevention of exposure is the basic rule for an emergency response. When an accident occurs that may generate an aerosol or droplets of infectious materials, particularly if the material is an agent requiring Biosafety Level 2 (or higher) precautions, the room should be evacuated immediately, the doors closed, and all clothing decontaminated. Enough time should be allowed for the droplets to settle and the aerosols to be reduced by the air changes of the ventilation system before attempting to decontaminate the area. The time required will depend upon the ventilation within the area, but a general rule is to wait approximately 30 minutes before reentry for decontamination. Protective clothing and approved respiratory protection should be worn during the decontamination to prevent personal exposure to the infectious agents that were released. The biosafety officer should be consulted before cleanup is started, to ensure that proper techniques will be employed.

FIGURE 5.1 The cleanup of an accidental spill of biohazardous material is illustrated. A laboratory worker using protective clothing, gloves, and respiratory protection is cleaning up the spill with paper towels that have been soaked with a disinfectant. The proper emergency response for a particular spillage will depend on the volume of the spill and the infectious hazard of the material. Courtesy, National Institutes of Health.

A spill of biohazardous material within a biological safety cabinet requires a special response and cleanup procedure. Cleanup should be initiated at once, while the cabinet continues to operate, using an effective chemical decontaminating agent (see Table 4.1). Aerosol generation during decontamination, and the escape of contaminants from the cabinet, should be prevented. Caution must be exercised in the choice of decontaminant, keeping in mind that fumes from flammable organic solvents, such as alcohol, can reach dangerous concentrations within a biological safety cabinet.

The proper emergency response for an accidental spillage of biohazardous material in the laboratory, outside a biological safety cabinet, will depend upon the hazard of the material and the volume. A minimally hazardous material that is spilled without generating significant aerosol may be cleaned up with a paper towel soaked with an effective decontaminating agent. A spill of a large volume of infectious material with the generation of aerosols will require cleanup personnel wearing protective clothing and respiratory protection (Figure 5.1). With *M. tuberculosis*, for example, the risk of exposure from the spill of a small quantity might be many times that of a much larger spill of *E. coli*. Therefore, if the agent is known, the recommended procedure and protective equipment should be used. Waiting approximately 30 minutes for the aerosols to settle before the cleanup of a large spill is essential. A spill kit or the best utensils available should be used to clean up, and material to be discarded should be placed in containers for decontamination and safe handling by others. Following cleanup, personnel should wash or shower.

Other types of spills that may generate hazardous aerosols include spills within centrifuges and the release of biohazardous materials within refrigerators, incubators, or shaker baths. The same principles discussed above apply: the area should be left immediately, protective equipment should be worn, the spill should be cleaned up, and the area should be disinfected. The personnel should then wash or shower.

As with biological spills, the proper emergency response to a chemical or mixed chemical/biological release will depend upon the hazard of the chemical and biological agents, the volume of material, and the location of the incident. The spill should be confined to a small area while avoiding the airborne release to the extent possible. The spill should be neutralized or flushed with water and followed with a cleanup or mopping up, with careful disposal of the residue. If the spilled material is highly volatile and noninfectious, it should be allowed to evaporate and be exhausted by the hood or ventilation system.

b. Handling of Spilled Solids

Generally, spilled solids of low toxicity should be swept into a dust pan and placed in an appropriate container for disposal. Additional precautions, such as the use of a vacuum cleaner equipped with a HEPA filter, may be necessary when cleaning up spills of more highly biohazardous materials.

c. Biological Radioactive Emergencies

The best way to avoid having a spill or other accident when working with radioactive materials is to preplan the work. Unfamiliar procedures should first be carried out without radioactive materials, so that problems will be discovered before the radioactive materials are utilized. Adequate time should be allotted for the experiment to prevent rushing at the end, as this can lead to clumsy or careless actions.

Should a spill occur, it is important to remember that spills of radioactive material are handled in a way similar to spills of infectious agents, except that there is additional concern for the radiation hazard. Determination of the primary hazard is of the utmost importance. In a spill involving both an infectious agent and a radioisotope, the radioactivity may be of secondary concern until the infectious agent has been inactivated. The disinfecting agent should be selected carefully: for example, hypochlorite will volatilize radioactive iodine.

The first concern in any spill, radioactive or otherwise, is to determine if anyone has been contaminated. Contaminated clothing should be removed immediately, and if there is a spill on the skin, the person should wash the contaminated area gently with mild soap and water. The laboratory should be evacuated unless the spill is contained within a hood.

The radiation safety officer (RSO) and supervisor should be notified immediately whenever there is a radioactive spill, regardless of its size. Laboratory personnel may be expected to clean up the spill, but the RSO is an essential resource and can provide important advice.

In all circumstances, the radioisotope and the approximate quantity spilled should be determined first. If the radioisotope is an energetic beta and/or gamma emitter, an external as well as internal hazard may exist. The external dose rates to individuals cleaning up the spill may be sufficient to require localized shielding and careful planning prior to cleanup. A significant external dose rate to the skin may result if the skin becomes contaminated during the cleanup. If the skin is damaged, an internal exposure may also occur. Depending on the chemical form, some radioisotopes may penetrate intact skin.

Examples of radioisotopes that would pose both external and internal hazards include, but are not limited to, ^{22}Na, ^{131}I, ^{32}P, and ^{36}Cl. An internal hazard may also exist if the radioactive material is volatile or is easily made airborne. Monitoring devices should be worn if they are normally required when handling the radioisotope in question, or if the RSO determines that they are necessary.

Low-energy beta emitters such as ^{14}C, ^{3}H, ^{35}S, and ^{45}Ca are usually not external hazards, provided that they are not deposited on the skin. Alpha particle emitters (^{241}Am, ^{239}Pu) are generally not considered external hazards but are very damaging if deposited inside of the body. For internal radiation hazards, the primary concern is to prevent the isotope from entering the body by the penetration of the skin. Appropriate protective equipment, such as gloves, should be worn during cleanup.

Prevention of skin contamination and of the generation of airborne contamination should be considered in all cases. Except for ^{3}H, the radioisotopes listed above can be detected with an appropriate Gieger counter or other monitoring equipment.

The spill should be cleaned up in a way that will minimize the generation of aerosols or the reentrainment of dusty materials. The area should be surrounded with absorbent material and cleaned from the outer edge inward to prevent increasing the size of the contaminated area.

All items used in cleaning up the spill should be disposed of as radioactive waste, to be decontaminated (if infectious). It is important not to sacrifice thoroughness in an effort to reduce the volume of the waste resulting from the cleanup.

Following cleanup, the area should be wipe-tested to verify that loose contamination has been removed. In addition, a Geiger counter survey will help find residual contamination. If residual contamination is found, the RSO should determine the requirements for additional cleanup. It may be easier to replace a floor tile than to spend hours scrubbing it to remove contamination.

It is important that the RSO be consulted, as it may be necessary to follow specific procedures described in the Nuclear Regulatory Commission license for the facility in question.

5. Other Emergencies

Laboratories should be prepared for problems resulting from severe weather or loss of a utility service. In the event of the latter, most ventilation systems not supplied with emergency power will become inoperative. All hazardous laboratory work should then cease until service has been restored and appropriate action has been taken to prevent exposure of personnel to hazardous or toxic agents.

F. REGULATION AND ACCREDITATION

The management, as well as those individuals directly responsible for the health and safety of employees, should be familiar with the statutory requirements of federal, state, and local governments that apply to the operations of their facility. In addition they should be aware of the accreditation processes and guidelines that may be available to assist the organization in complying with legal requirements. Numerous governmental agencies are involved in the regulatory process, and the regulations promulgated by them may be changed frequently, or new regulations may be issued. Different laboratories are affected by these requirements to different degrees depending on the number of employees, the types of hazardous materials handled, and the nature of its operation (e.g., manufacture, research, hospital support, or teaching). In addition to the need to comply, there frequently are requirements for recordkeeping to document adherence to the regulation. Some of the regulations provide for inspections to ascertain compliance. These inspections may be unannounced, or they can be initiated at the request of the employer or employees.

Large organizations, professional societies, and publishers of specialty newsletters monitor the *Federal Register*, a publication of the federal government that is used by the different agencies to publish proposed and final regulations. Professional societies usually keep their constituencies notified of pertinent matters through their newsletters. It is important for interested or affected parties to know when new regulations, or changes in existing ones, are being proposed, so that they can take advantage of the opportunity to participate in the regulatory process.

Three other useful sources of information are (1) the *Congressional Regulatory/Federal Regulatory Directory*, published by Congressional Quarterly, Inc., Washington, DC 20037; (2) the *United States Government Manual*, published by the U.S. Government Printing Office for the Office of the Federal Register, National Archives and Record Administration, Washington, DC 20402; and (3) the *Occupational Safety and Health Reporter*, Bureau of National Affairs, Inc., Washington, DC 20037. The first two publications identify specific offices and provide telephone numbers for obtaining additional information about regulations.

A glossary of regulatory definitions, as well as lists of regulatory agencies and accrediting bodies, is provided in Appendix G.

G. TEACHING BIOSAFETY IN ACADEMIC SETTINGS

1. Introduction

An academic biosafety training program is based on the size of the institution and the relative biohazards found there. In cases where institutional safety personnel are not available or are not themselves trained in biosafety, a faculty member with expertise in microbiology may provide training in biosafety as an additional academic responsibility. If the program is large enough, a biosafety officer may be employed to provide the training as well as advice on containment levels and practices.

2. Safety in Laboratory Courses

It is essential that safe practices be taught in academic courses at all levels, to prepare students for future responsibilities as principal investigators, teachers, medical/project directors, or supervisors of laboratories. The management of teaching laboratories in academic institutions may be delegated to a technician or to an academician, but the responsibility for safety management still resides with the department chairperson.

3. Orientation and Training of Students

Safety training should begin with an orientation session in which general safety policies and a positive attitude toward safety are introduced. Written institutional safety policies should be discussed and the seriousness of biosafety impressed upon the students. Safety should also be incorporated into lectures, seminars, audiovisual presentations, poster sessions, laboratory exercises, and other aspects of the academic experience. There should be training in the use of safety aids for controlling specific hazards. Biosafety training should be a required part of the core curriculum: students should not graduate without being able to handle safely the hazardous agents in their chosen field. Safe procedures and practices should be understood and mastered by all who are to work with biohazardous agents in unsupervised advanced laboratory exercises, in special research projects, and even in unsupervised hospital laboratories (so-called because of their use by interns and residents in medical or nursing programs).

Medical students, nursing students, and other personnel need to be taught to work safely in such laboratories. They should not handle blood or body fluid specimens or cultures of microorganisms without training in standard operating procedures. These individuals also need training in the use of infection control "isolation" practices when dealing with patients who have certain contagious diseases or increased susceptibility to infection. Infectious materials from patients on "isolation precautions" should not be processed in the hospital laboratories unless appropriate containment equipment is available and specific training has been received. Students should learn to adhere to the personal practices required for the work, especially the restrictions on eating, drinking, and smoking in all laboratories. Hospital laboratories should be operated under the Biosafety Level 2 conditions recommended by the CDC/NIH guidelines (Appendix A).

Graduate students or advanced undergraduate students working with biohazardous materials should be as knowledgeable about safety practices as any certified medical technician. Training in the procedures, equipment, and facilities necessary for each biosafety level should be a part of the knowledge base required for a graduate degree in microbiology and in other fields that involve the handling of infectious agents. The principal investigator or laboratory instructor should ensure that the student knows the hazards of the work, as well as the appropriate containment, personal practices, and equipment to be used. For example, such students should learn how to use an autoclave or chemical disinfectant to effectively decontaminate biohazardous laboratory waste, and should learn how to monitor the process. Students also should be given information about the specific agents to be handled, including modes of transmission, symptoms of disease, and risk factors. The student should sign a document to acknowledge that training and information have been provided. If immunizations are recommended for work with the agent, vaccines should be provided along with appropriate medical surveillance and access to medical care.

Academic institutions should provide biosafety training for their assistants who teach in such laboratory courses as microbiology, immunology, biochemistry, and molecular biology. A written list of safety procedures and precautions should be tailored for each laboratory exercise, according to the materials being used and the level of hazard.

4. Design of Safe Laboratory Exercises and Experiments

The safety guidelines and regulations of local, state, and federal agencies (e.g., the CDC/NIH Guidelines, Appendix A) are also applicable to teaching laboratories that work with biohazardous materials. A biosafety officer or a knowledgeable faculty member should review the procedures to be carried out as laboratory exercises and determine if the containment practices need to be improved to reduce the risk. For example, some laboratory exercises might be miniaturized, reducing the level of risk by reducing the volume or potential dose of the infectious agent. In other exercises, attenuated microbial strains might be used, or a nonpathogenic organism such as

Bacillus subtilis might be substituted for a pathogen in demonstrating a routine technique such as the streaking of a plate. Pathogens such as *Salmonella typhi* or *Brucella suis*, which have caused laboratory infections in the past or which are known to be highly infectious, should be used in sealed demonstration plates and tubes rather than in "hands-on" procedures. All directors of teaching and training laboratories that use biohazardous agents are urged to review their procedures in order to minimize the risk to their students and trainees.

5. Monitoring and Recordkeeping

The institutional legal office should be consulted about recordkeeping requirements, such as the OSHA log of occupational injuries and illnesses, which may be required by law. In order to correct problems, investigations of accidents and "near misses" should be documented. It is important that accidents be investigated in a timely manner, and that accident reports be completed by the teaching assistant or laboratory supervisor and forwarded promptly to the appropriate individuals for investigation. The reports may need to be confirmed by a responsible departmental representative. The person submitting a report should include an assessment of how the accident could have been prevented. The university legal office should provide advice as to the recordkeeping requirements for injuries and accidents involving students.

References

References for the report are listed here in alphabetical order by first author. Each reference is cross-indexed to where it is cited in the text by chapter number, section letter, and page number. This informaton appears in square brackets at the end of each reference.

1. Advisory Committee on Dangerous Pathogens: Acquired Immunodeficiency Syndrome (AIDS). 1984. Interim Guidelines. London: Health and Safety Executive. [3, E, p. 17]

2. Advisory Committee on Immunization, Council of Medical Sciences. 1985. Guide for Adult Immunizations. Philadelphia, PA: American College of Physicians. [5, D, p. 58]

3. Alexander MT and Brandon BA. 1986. Packaging and Shipping of Biological Materials at the ATCC. Rockville, MD: American Type Culture Collection. [3, G, p. 21]

4. American Industrial Hygiene Association Biohazards Committee. 1985. Biohazards Reference Manual. Akron, OH: American Industrial Hygiene Association. [Preface, p. vii], [3, K, pp. 30,32]

5. American Industrial Hygiene Association Biohazards Committee. 1985. Allergies. Pp. 13-19 in Biohazards Reference Manual. Akron, OH: American Industrial Hygiene Association. [5, D, p. 57]

6. American Industrial Hygiene Association Biohazards Committee. 1985. Pp. 128-136 in Biohazards Reference Manual. Akron, OH: American Industrial Hygiene Association [3, C, p. 14] [Appendix D]

7. American Occupational Medical Association. 1976. Code of Ethical Conduct for Physicians Providing Occupational Medical Services. 55 West Seegers Road, Arlington Heights, IL 60005. [5, D, p. 59]

8. Anonymous. 1974. Report of the Committee of Inquiry into the Smallpox Outbreak in March and April 1973. London: Her Majesty's Stationery Office. [3, F, p. 18]

9. Asher DM, Gibbs CJ Jr, and Gajdusek DC. 1986. Slow viral infections: Safe handling of the agents of subacute spongiform encephalopathies. Pp. 59-71 in Laboratory Safety: Principles and Practices, BM Miller, DHM Groschel, JH Richardson, D Vesley, JR Songer, RD Housewright, and WE Barkley, eds. Washington, DC: American Society for Microbiology. [3, E, pp. 16,17]

10. Bauer M and Patnode R. 1984. Health Hazard Evaluation Report. HETA 81-121-1421. Insect Rearing Facilities. Agricultural Research Service, U.S. Department of Agriculture. National Institute for Occupational Safety and Health. [3, C, p. 15]

11. Bellas TE. 1982. Insects as a Cause of Inhalant Allergies: A Bibliography. Division of Entomology Report No. 25. Canberra City, Australia: Commonwealth Scientific and Industrial Research Organization. [3, C, p. 15]

12. Belssey RE, Baer DM, Statland BE, and Sewell DL. 1986. The Physicians' Laboratory. Brooklyn, NY: Medical Economics Books. [3, L, p. 33]

13. Bitton G. 1980. Introduction to Environmental Virology. P. 48. New York: John Wiley and Sons. [3, K, p. 30]

14. Bland SM, Evans IR, and Rivera JC. 1987. Allergy to laboratory animals in health care personnel. Occupational Medicine 2:525-546. [3, C, p. 15], [5, D, p. 57]

15. Blaser MJ and Lofgren JP. 1982. Fatal salmonellosis originating in a clinical microbiology laboratory. Journal of Clinical Microbiology 13(5):855-858. [3, F, p. 18]

16. Bleckman J. 1986. Incinerator Maintenance Requirements. American Hospital Association Technical Document Series No. 055883. Chicago, IL: American Society for Hospital Engineering, American Hospital Association. [4, E, p. 41]

17. Blenden DC and Adldinger HK. 1986. Transmission and control of viral zoonoses in the laboratory. Pp. 72-89 in Laboratory Safety: Principles and Practices, BM Miller, DHM Groschel, JH Richardson, D Vesley, JR Songer, RD Housewright, and WE Barkley, eds. Washington, DC: American Society for Microbiology. [3, C, p. 14] [Appendix D]

18. Block SS. 1983. Disinfection, Sterilization and Preservation. 3rd edition. Philadelphia, PA: Lea and Febiger. [3, K, p. 30]

19. Bloom HM. 1986. Designs to simplify laboratory construction and maintenance, improve safety, and conserve energy. Pp. 138-143 in Laboratory Safety: Principles and Practices, BM Miller, DHM Groschel, JH Richardson, D Vesley, JR Songer, RD Housewright, and WE Barkley, eds. Washington, DC: American Society for Microbiology. [5, B, p. 48]

20. Bond WW, Favero MS, Peterson NJ, Gravelle CR, Ebert JW, and Maynard JE. 1981. Survival of hepatitis B virus after drying and storage for one week. Lancet I:550-551. [2, C, p. 11]

21. Boyce JM and Kaufmann AF. 1986. Transmission of bacterial and rickettsial zoonoses in the laboratory. Pp. 90-99 in Laboratory Safety: Principles and Practices, BM Miller, DHM Groschel, JH Richardson, D Vesley, JR Songer, RD Housewright, and WE Barkley, eds. Washington, DC: American Society for Microbiology. [3, C, p. 15] [Appendix D]

22. Bretherick L. 1985. Handbook of Reactive Chemical Hazards. 3rd edition. Boston, MA: Butterworths. [4, F, p. 43]

23. Brown P, Gibbs CJ Jr, Amyx HL, Kingsbury DT, Rowher RG, Sulima MP, and Gajdusek DC. 1982. Chemical disinfection of Creutzfeldt-Jakob disease virus. New England Journal of Medicine 306, 21:1279-1282. [2, C, p. 12]

24. Brown P, Rohwer RG, and Gajdusek DC. 1986. Concise communications: Newer data on the inactivation of scrapie virus or Creutzfeldt-Jakob disease virus in brain tissue. Journal of Infectious Diseases 156:6:1145-1148. [3, E, pp. 16,17]

25. Brunner CR. 1984. Incineration Systems: Selection and Design. New York, NY: Van Nostrand Reinhold. [4, E, p. 41]

26. Carden GA, Juranek DD, and Melvin DM. 1986. Laboratory acquired parasitic infections. Pp. 100-111 in Laboratory Safety: Principles and Practices, BM Miller, DHM Groschel, JH Richardson, D Vesley, JR Songer, RD Housewright and WE Barkley, eds. Washington, DC: American Society for Microbiology. [3, C, p. 14] [Appendix D]

27. Center for Disease Control. 1972. Human rabies—Texas. Morbidity and Mortality Weekly Report 21(14):113-114. [2, C, p. 11]

28. Center for Disease Control. 1977. Clinical and Public Health Laboratory Survey, 1976. Laboratory Program Office. Atlanta, GA: Center for Disease Control. [2, B, p. 8]

29. Center for Disease Control. 1977. Rabies in a laboratory worker—New York. Morbidity and Mortality Weekly Report 26(22):183-184. [2, C, p. 11]

30. Centers for Disease Control. 1985. Annual summary 1984. Reported morbidity and mortality in the United States. Morbidity and Mortality Weekly Report 33:54. [2, C, p. 9]

31. Centers for Disease Control. 1985. Hepatitis surveillance 1982-83. Morbidity and Mortality Weekly Report 34(1SS):1ss-10ss. [2, B, pp. 9,10]

32. Centers for Disease Control. 1986. Tuberculosis and acquired immunodeficiency syndrome, Florida. Morbidity and Mortality Weekly Report 35:587-90. [2, C, p. 12]

33. Centers for Disease Control. 1986. Tuberculosis—United States, 1985, and the possible impact of human T-lymphotropic virus type III/lymphadenopathy-associated virus infections. Morbidity and Mortality Weekly Report 35:64-66. [2, C, p. 10]

34. Centers for Disease Control. 1987. Recommendations for prevention of HIV transmission in health-care settings. Morbidity and Mortality Weekly Report 36(2S):3S-18S. [1, A, p. 2], [2, C, p. 12], [3, A, p. 13], [3, H, p. 22], [Appendix C]

35. Centers for Disease Control. 1987. Update: Human immunodeficiency virus infections in health-care workers exposed to blood of infected patients. Morbidity and Mortality Weekly Report 36:285-289. [2, C, p. 12]

36. Centers for Disease Control. 1988. Agent summary statement for human immunodeficiency viruses (HIVs) including HTLV-III, LAV, HIV-1 and HIV-2. Morbidity and Mortality Weekly Report 37(No. S-4):1-17. [2, C, pp. 10,12], [3, A, p. 13], [Appendix B]

37. Centers for Disease Control. 1988. Occupationally acquired human immunodeficiency virus infections in laboratories producing virus concentrates in large quantities: Conclusions and recommendations of an expert team convened by the Director of the National Institutes of Health (NIH). Morbidity and Mortality Weekly Report 37(No. S-4):19-22. [3, B, p. 13]

38. Centers for Disease Control. 1988. Update: Universal precautions for prevention of transmission of human immunodeficiency virus, hepatitis B virus, and other bloodborne pathogens in health-care settings. Morbidity and Mortality Weekly Report 37(24):377+. [1, A, p. 2], [2, C, p. 12], [3, A, p. 13], [3, H, p. 22], [Appendix C]

39. Chanlett ET. 1980. Solid waste disposal. Pp. 950-974 in Maxcy-Rosenau Public Health and Preventive Medicine, JM Last, ed. New York, NY: Appleton-Century-Crofts. [4, A, pp. 34,35]

40. Chick EW, Bauman DS, Lapp NL, and Morgan WKC. 1972. A combined field and laboratory epidemic of histoplasmosis: Isolation from bat feces in West Virginia. American Reviews of Respiratory Disease 105:968-971. [3, F, p. 18]

41. College of American Pathologists. 1987. Standards for Laboratory Accreditation. Skokie, IL: College of American Pathologists. [5, B, p. 48]

42. Collins, CH. 1983. Laboratory-Acquired Infections: Incidence, Causes, and Prevention. London: Butterworths. [2, C, p. 11], [3, B, p. 13]

43. Communicable Disease Center. 1964. Zoonoses Surveillance: Brucellosis Annual Summary, 1963. Department of Health, Education, and Welfare, U.S. Public Health Service. Atlanta, GA: Communicable Disease Center. [2, C, p. 9]

44. Council on Accreditation. 1987. Guide for the Care and Use of Laboratory Animals. Joliet, IL: American Association for Accreditation of Laboratory Animal Care. [5, C, p. 54]

45. Dandoy SE, Kirkman-Liff BL, and Krakowski FM. 1984. Hepatitis B exposure incidents in community hospitals. American Journal of Public Health 74(8):804-807. [2, B, p. 9], [2, C, pp. 11]

46. Davidson WL and Hummler K. 1960. B-virus infection in man. Annals of the New York Academy of Science 85:9970-979. [3, D, p. 15]

47. Department of Health and Social Security. Scottish Home and Health Department. Department of Health and Social Services Northern Ireland. Welsh Office. Code of Practice for the Prevention of Infection in Clinical Laboratories and Postmortem Rooms. London: Her Majesty's Stationery Office. [3, E, p. 16]

48. Disinfectants. Pp. 56-57 in Official Methods of Analysis of the Association of Analytical Chemists, W Horowitz, ed., 13th edition. Washington, DC: Association of Official Analytical Chemists. [4, E, p. 42]

49. Doucet LG. 1986. Controlled Air Incineration: Design, Procurement, and Operational Considerations. American Hospital Association Technical Document Series. Chicago, IL: American Society for Hospital Engineering, American Hospital Association. [4, E, pp. 41,42]

50. Edlich RF, Levesque E, Morgan RF, Kenney JG, Silloway KA, and Thacker JG. 1986. Laboratory personnel as the first responders. Pp. 279-291 in Laboratory Safety: Principles and Practices, BM Miller, DHM Groschel, JH Richardson, D Vesley, JR Songer, RD Housewright, and WE Barkley, eds. Washington, DC: American Society for Microbiology. [5, D, p. 59]

51. Elliott LJ, Carson G, Pauker S, West D, Wallingford K, and Griefe A. 1983. Industrial Hygiene Characterization of Commercial Applications of Genetic Engineering and Biotechnology. Cincinnati, OH: National Institute for Occupational Safety and Health, Division of Surveillance, Hazard Evaluations and Field Studies. [3, K, pp. 30,31,32]

52. Favero MS. 1983. Chemical disinfection of medical and surgical materials. Pp. 469-492 in Disinfection, Sterilization, and Preservation, SS Block, ed. Philadelphia, PA: Lea and Febiger. [4, E, p. 42]

53. Favero MS, Petersen NJ, and Bon WW. 1986. Transmission and control of laboratory acquired hepatitis infection. Pp. 49-58 in Laboratory Safety: Principles and Practices, BM Miller, DHM Groschel, JH Richardson, D Vesley, JR Songer, RD Housewright, and WE Barkley, eds. Washington, DC: American Society for Microbiology. [3, C, p. 14] [Appendix B] [Appendix D]

54. Fischer PM, Addison LA, Curtis P, and Mitchell JM. 1983. The Office Laboratory. East Norwalk, CT: Appleton-Century-Crofts. [3, L, p. 33]

55. Gajdusek DC, Gibbs CJ Jr, Asher MD, Brown P, Diwan A, Hoffman P, Nemo G, Rohwer R, and White L. 1977. Precautions in medical care of, and in handling materials from, patients with transmissible virus dementia (Creutzfeldt-Jakob Disease). New England Journal of Medicine 297:1253-1258. [3, E, p. 16]

56. Gajdusek DC, Gibbs CJ Jr, Collins G, and Traub RD. 1976. Survival of Creutzfeldt-Jakob disease virus in formal-fixed brain tissue. New England Journal of Medicine 294:553. [3, E, pp. 16,17]

57. Goldman RH. 1986. General occupational health history and examination. Journal of Occupational Medicine 28:967-974. [5, D, pp. 55,57]

58. Goldman RH and Peters JM. 1981. The occupational and environmental history. Journal of the American Medical Association 246:2831-2836. [5, D, pp. 55,57]

59. Grieble HG, Bird TJ, Nidea HM, and Miller CA. 1974. Chute-hydropulping waste disposal system: A reservoir of enteric bacilli Pseudomonas in a modern hospital. Journal of Infectious Diseases 130:602-607. [4, B, p. 36]

60. Grist NR and Emslie J. 1985. Infections in British clinical laboratories, 1982-3. Journal of Clinical Pathology 38:721-725. [3, B, p. 13]

61. Groschel DHM. Safety in clinical microbiology laboratories. Pp. 32-35 in Laboratory Safety: Principles and Practices, BM Miller, DHM Groschel, JH Richardson, D Vesley, JR Songer, RD Housewright, and WE Barkley, eds. Washington, DC: American Society for Microbiology. [5, C, p. 52]

62. Halperin WE and Frazier TM. 1985. Surveillance for the effect of workplace exposure. Annual Reviews of Public Health 6:419-432. [5, D, pp. 55,58]

63. Hanel E and Halbert MM. 1986. Pipetting. Pp. 204-214 in Laboratory Safety: Principles and Practices, BM Miller, DHM Groschel, JH Richardson, D Vesley, JR Songer, RD Housewright, and WE Barkley, eds. Washington, DC: American Society for Microbiology. [3, I, p. 23], [3, J, p. 26]

64. Harrington JM and Shannon HS. 1976. Incidence of tuberculosis, hepatitis, brucellosis, and shigellosis in British medical laboratory workers. British Medical Journal 1:759-762. [2, C, pp. 10,11]

65. Harris-Smith R and Evans CGT. 1968. The Porton mobile enclosed chemostat (POMEC). In Continuous Cultivation of Micro-organisms. Proceedings of 4th Symposium. Prague, Czechoslovakia: Academa. [3, K, p. 30]

66. Hicks CG, Hargiss CO, and Harris JR. 1985. Prevalence survey for hepatitis B in high risk university employees. American Journal of Infection Control 13:1-6. [2, C, p. 10], [3, H, p. 22]

67. Hinman AR, Fraser DW, Douglas RG, Bowen GS, Kraus AL, Winkler WG, and Rhodes WW. 1975. Outbreak of lymphocytic choriomeningitis virus infections in medical center personnel. American Journal of Epidemiology 101(2):103-110. [3, F, p. 18]

68. Hoogstral H, Meegan JM, Khabil GM, and Adham FK. 1979. The Rift Valley fever enzootic in Egypt 1977-1978: 2. Ecological and entomological studies. Transactions of the Royal Society of Tropical Medicine and Hygiene 73(6):624-629. [3, F, p. 18]

69. Hubbert WT, McCulloch WF, and Schnurrenberger PR, eds. 1975. Diseases Transmitted from Animals to Man. Springfield, IL: Charles C Thomas. [3, C, p. 14] [Appendix D]

70. Jacobsen JT, Orlob RB, and Clayton JL. 1985. Infections acquired in clinical laboratories in Utah. Journal of Clinical Microbiology 21(4):486-489. [2, C, p. 10], [3, B, p. 13]

71. Jaugstetter JE and Wagner WM. 1986. Packaging and shipping biological, radioactive and chemical materials. Pp. 215-227 in Laboratory Safety: Principles and Practices, BM Miller, DHM Groschel, JH Richardson, D Vesley, JR Songer, RD Housewright, and WE Barkley, eds. Washington, DC: American Society for Microbiology. [3, G, p. 20]

72. Joint Commission on Accreditation of Healthcare Organizations. 1988. AMH/87 Accreditation Manual for Hospitals. [ISPNO-86688-104-2]. Chicago, IL: Joint Commission on Accreditation of Healthcare Organizations. [5, B, p. 48]

73. Kent PS and Kubica GP. 1985. A Guide to the Level III Laboratory. U.S. Department of Health and Human Services, U.S. Public Health Service, Centers for Disease Control. Atlanta, GA: Centers for Disease Control. [3, J, p. 26]

74. Kubica GP. 1987. Personal communication. [2, C, p. 10]

75. Lauer JL, Battles DR, and Vesley D. 1982. Decontaminating infectious laboratory waste by autoclaving. Applied Environmental Microbiology 44:690-694. [4, E, p. 42]

76. Lauer JL, Van Drunen NA, Washburn JW, and Balfour HH. 1979. Transmission of hepatitis B virus in clinical laboratory areas. Journal of Infectious Diseases 140:513-516. [2, C, p. 11]

77. Lee HW and Johnson KM. 1982. Laboratory-acquired infections with Hantaan virus, the etiologic agent of Korean hemorrhagic fever. Journal of Infectious Diseases 146(5):645-651. [3, F, p. 18]

78. MacArthur S, Jacobson R, Marrero H, Rahman Z, and Schneiderman H. 1986. Autopsy removal of the brain in AIDS: a new technique (letter). Human Pathology 17(12):1296-1297. [3, E, p. 17]

79. MacArthur S and Schneiderman H. 1987. Infection control and the autopsy of persons with human immunodeficiency virus. American Journal of Infection Control 15:4:172-177. [3, E, pp. 16,17], [5, B, p. 48]

80. Masters CL, Jacobsen P, and Kakulas B. 1985. Letter to the Editor. Journal of Neuropathology and Experimental Neurology 44:3:304-306. [3, E, p. 17]

81. McCormick RD and Maki DG. 1981. Epidemiology of needle-stick injuries in hospital personnel. American Journal of Medicine 70:928-932. [2, C, pp. 10,11]

82. Medical Research Council of Canada. 1980. Guidelines for the Handling of Recombinant DNA Molecules and Animal Viruses and Cells. Ottawa. Cat. No. MR21-1/1977, ISBN 0-662-00587-2. [3, K, p. 30]

83. Miller BM, Groschel DHM, Richardson JH, Vesley D, Songer JR, Housewright RD, and Barkley WE, eds. 1986. Laboratory Safety: Principles and Practices. Washington, DC: American Society for Microbiology. [Preface, p. vii], [Table 4.1], [5, E, p. 63]

84. Miller CD, Songer JR, and Sullivan JF. 1987. A twenty-five year review of laboratory-acquired human infections at the National Animal Disease Center. American Industrial Hygiene Association 48:271-275. [2, C, p. 10], [3, B, p. 13], [3, F, p. 18]

85. Miller DC. 1988. Creutzfeldt-Jakob disease in histopathology technicians. Letter to the Editor. New England Journal of Medicine 318:853-854. [2, C, p. 12]

86. Miller JM. 1986. Receiving and handling biological specimens safely. Pp. 57-64 in Proceedings of the 1985 Institute on Critical Issues in Health Laboratory Practice: Safety Management in the Public Health Laboratory, JH Richardson, E Schoenfeld, JJ Tulis, and WM Wagner, eds. Wilmington, DE: E.I. du Pont de Nemours and Company, Inc. [3, G. pp. 20,21]

87. Mitscherlich E and Marth EH. 1984. Bacteria and rickettsiae important in human and animal health. In Microbial Survival in the Environment. New York, NY: Springer-Verlag. [4, A, p. 35]

88. National Committee for Clinical Laboratory Standards (NCCLS). 1985. Preparation and Testing of Reagent Water in the Clinical Laboratory. Second edition. Proposed Guidelines 5:315-363. NCCLS Document C3-P2. Villanova, PA: NCCLS. [5, B, p. 50]

89. National Committee for Clinical Laboratory Standards (NCCLS). 1986. Clinical Laboratory Hazardous Waste. Proposed Guidelines 6:395-474. NCCLS Document GPS-P. Villanova, PA: NCCLS. [5, B, p. 50]

90. National Committee for Clinical Laboratory Standards (NCCLS). 1987. Protection of Laboratory Workers from Infectious Disease Transmitted by Blood and Tissue. Proposed Guidelines 7(9):325-431. NCCLS Document M29-P. Villanova, PA: NCCLS. [1, A, p. 2], [3, E, p. 16]

91. National Fire Protection Association. 1978. Fire Protection Guide on Hazardous Materials. 7th edition. Quincy, MA: National Fire Protection Association. [4, F, p. 43]

92. National Fire Protection Association. 1987. Laboratories in Health Related Institutions. NFPA No. 56C. Boston, MA: National Fire Protection Association. [5, B, p. 48]

93. National Research Council, Committee on Hazardous Substances in the Laboratory. 1981. Prudent Practices for Handling Hazardous Chemicals in Laboratories. Washington, DC: National Academy Press. [Preface, p. vii]

94. National Research Council, Committee on Hazardous Substances in the Laboratory. 1983. Prudent Practices for Disposal of Chemicals from Laboratories. Washington, DC: National Academy Press. [Preface, p. vii]

95. National Sanitation Foundation. 1983. Appendix E: Standard 49 for Class II (laminar flow) biohazard cabinetry. In Recommended Microbiological Decontamination Procedure. Ann Arbor, MI: National Sanitation Foundation. [4, E, p. 39], [5, B, p. 50]

96. Nelson KE, Rubin FL, and Andersen B. 1975. An unusual outbreak of brucellosis. Archives of Internal Medicine 135:691-695. [3, F, p. 18]

97. Osterholm M and Andrews JS. 1979. Viral hepatitis in hospital personnel in Minnesota. Report of a statewide survey. Minnesota Medicine September, 683-689. [2, C, p. 10]

98. Phillips GB. 1969. Control of microbiological hazards in the laboratory. American Industrial Hygiene Association Journal 30:170-176. [3, I, p. 23]

99. Phillips GB and Baily SP. 1966. Hazards of mouth pipetting. American Journal of Medical Technology 32:127-129. [3, I, p. 23]

100. Pike RM. 1976. Laboratory-associated infections. Summary and analysis of 3,921 cases. Health Laboratory Science. 13:105-114. [2, A, p. 8], [2, C, pp. 8,9,11,12], [3, I, p. 23], [3, J, p. 26], [3, K, p. 28]

101. Pike RM. 1979. Laboratory-associated infections: Incidence, fatalities, causes, and prevention. Annual Reviews of Microbiology 33:41-66. [3, B, p. 13], [3, F, pp. 18,20]

102. Pike RM, Sulkin SE, and Schulze ML. 1965. Continuing importance of laboratory-acquired infections. American Journal of Public Health 55:190-199. [3, I, p. 23]

103. Reeves SA and Noble RC. 1985. Ineffectiveness of annual chest roentgenograms in tuberculin skin-test positive hospital employees. American Journal of Infection Control 11:212-216. [5, D, p. 58]

104. Reitman M and Phillips GB. 1955. Biological hazards of common laboratory procedures. I. The pipette. American Journal of Medical Technology 21:338-342. [3, I, p. 23]

105. Richardson JH and Barkley WE, eds. 1984. Biosafety in Microbiological and Biomedical Laboratories. U.S. Public Health Service. Centers for Disease Control and National Institutes of Health. HHS Publication No. (CDC) 84-8395. Washington, DC: U.S. Government Printing Office. [2, B, p. 9], [2, C, p. 12], [3, A, p. 13], [3, B, p. 13], [3, G, p. 21], [3, K, pp. 30], [5, C, p. 54], [5, D, pp. 59,62], [Appendix B]

106. Richardson JH, Schoenfeld E, Tulis JJ, and Wagner WM, eds. 1986. Proceedings of the 1985 Institute on Critical Issues in Health Laboratory Practice: Safety Management in the Public Health Laboratory. Wilmington, DE: E.I. du Pont de Nemours and Company, Inc. [5, B, p. 48], [5, C, p. 52]

107. Riley RL. 1957. Aerial dissemination of pulmonary tuberculosis. American Review of Tuberculosis 76:931-941. [3, F, p. 20]

108. Riley RL. 1961. Airborne pulmonary tuberculosis. Bacteriology Reviews 25:243-248. [2, F, p. 20]

109. Rosenberg RN, White CL, Brown P, Gajdusek DC, Volpe JJ, Posner J, and Dyck PJ. 1986. Precautions in handling tissues, fluids, and other contaminated materials from patients with documented or suspected Creutzfeldt-Jakob Disease. Committee on Health Care Issues, American Neurological Association. Annals of Neurology 19:75-77. [3, E, p. 16]

110. Rubin J. 1977. Tuberculocidal agents. P. 422 in Disinfection, Sterilization and Preservation, SS Block, ed. Philadelphia, PA: Lea and Febiger. [3, E, p. 18]

111. Rutala WA, Steigel MM, and Smith FA, Jr. 1982. Decontamination of laboratory microbiological waste by steam sterilization. Applied Environmental Microbiology 43:1311-1316. [4, E, p. 42]

112. Sherertz RJ and Hampton AL. 1986. Infection control aspects of hospital employee health. In Prevention and Control of Nosocomial Infections, RP Wenzel, ed. Baltimore, MD: Williams and Wilkins. [5, D, pp. 56,58]

113. Sitwell L, Lach B, Atack E, and Atack D. 1988. Letter to the Editor. New England Journal of Medicine 318:854. [2, C, p. 12]

114. Skinhoj P. 1974. Occupational risks in Danish clinical chemistry laboratories. II. Infections. Scandanavian Journal of Clinical Laboratory Investigation 33:27-29. [2, C, p. 10]

115. Smith GE, Simpson DTH, Brown TW, and Zlotnik I. 1967. Fatal human disease from Vervet monkeys. Lancet 2(256):1129-1130. [3, D, p. 15]

116. Sobsey MD. 1982. Detection methods for viruses in solid waste landfill leachates. Pp. 171-178 in Methods in Environmental Virology, CP Gerba and SM Goya, eds. New York, NY: Marcel Dekker. [4, A, p. 35]

117. Songer JR. 1986. Safety management: Laboratory safety program organization. Pp. 1-9 in Laboratory Safety: Principles and Practices. BM Miller, DHM Groschel, JH Richardson, D Vesley, JR Songer, RD Housewright, and WE Barkley, eds. Washington, DC: American Society for Microbiology. [5, C, p. 52]

118. Spyker DA. 1986. Allergic reactions and poisoning. Pp. 267-268 in Laboratory Safety: Principles and Practices. BM Miller, DHM Groschel, JH Richardson, D Vesley, JR Songer, RD Housewright, and WE Barkley, eds. Washington, DC: American Society for Microbiology. [3, K, p. 28]

119. Stoneburner RL and Kristal A. 1985. Increasing tuberculosis incidence and its relationship to acquired immunodeficiency syndrome in New York City. Presented at the International Conference on the Acquired Immunodeficiency Syndrome (AIDS), Atlanta, GA. April 1985. [2, C, p. 10]

120. Sulkin SE and Pike RM. 1951. Survey of laboratory-acquired infections. American Journal of Public Health 41:769-781. [3, B, p. 13], [3, I, p. 23]

121. Sullivan JF, Songer JR, and Estreme IE. 1976. Laboratory-acquired infections at the National Animal Diseases Center 1960-1976. Health Laboratory Science 15:58-64. [3, B, pp. 13]

122. Sunderam G, McDonald RJ, Maniatis T, Oleske J, Kapila R, and Reichman LB. 1986. Tuberculosis as a manifestation of the acquired immunodeficiency syndrome (AIDS). Journal of the American Medical Association 256:362-366. [2, C, p. 10]

123. Swisher BL and Ewing EP Jr. 1986. Frozen section technique for tissues infected by the AIDS virus. Journal of Histotechnology 9:29. [3, E, p. 18]

124. U.S. Code of Federal Regulations, Title 7, Department of Agriculture, Parts 71-123. [3, K, p. 30]

125. U.S. Code of Federal Regulations, Title 9, Department of Agriculture, Part 122, Section 122.1-122.4. [3, G, p. 21]

126. U.S. Code of Federal Regulations, Title 29, Occupational Safety and Health Administration, Part 1910.1200. [5, A, p. 48], [5, C, p. 52]

127. U.S. Code of Federal Regulations, Title 40, Environmental Protection Agency, Parts 300-330. [5, E, p. 63]

128. U.S. Code of Federal Regulations, Title 42, Public Health Service, Part 71.156. [3, G, p. 21]

129. U.S. Code of Federal Regulations, Title 42, Public Health Service, Part 72. [3, G, p. 20], [3, K, p. 30]

130. U.S. Code of Federal Regulations, Title 49, Department of Transportation, Parts 100-199. [3, K, p. 30]

131. U.S. Code of Federal Regulations, Title 49, Department of Transportation, Parts 173.386-173.387. [3, G, p. 21]

132. U.S. Department of Health, Education, and Welfare. 1974. Biological Safety Manual for Research Involving Oncogenic Viruses. Part I: NCI Safety Standards for Research Involving Oncogenic Viruses. National Cancer Institute, Office of Research Safety. DHEW Publication No. (NIH) 76-116. Washington, DC: U.S. Government Printing Office. [3, K, p. 30]

133. U.S. Department of Health, Education and Welfare. 1974. National Institutes of Health Biohazards Safety Guide. U.S. Public Health Service, National Institutes of Health. Washington, DC: U.S. Government Printing Office. Stock No. 1740-00383. [3, K, p. 30]

134. U.S. Department of Health and Human Services. 1979. Laboratory Safety Monograph—A Supplement to the NIH Guidelines for Recombinant DNA Research. U.S. Public Health Service, National Institutes of Health. Bethesda, MD: National Institutes of Health. [1, A, p. 1], [3, K, p. 30], [5, D, p. 57]

135. U.S. Department of Health and Human Services. 1982. Medical Surveillance of Biotechnology Workers. Report of the CDC/NIOSH Ad Hoc Working Group on Medical Surveillance for Industrial Applications of Recombinant DNA. U.S. Public Health Service, Centers for Disease Control, National Institute for Occupational Safety and Health. Recombinant DNA Technical Bulletin 5:133-138. [3, K, p. 30], [5, D, p. 55]

136. U.S. Department of Health and Human Services. 1986. Guidelines for Research Involving Recombinant DNA Molecules. U.S. Public Health Service, National Institutes of Health. Federal Register 51(88):16957-16985. Bethesda, MD: National Institutes of Health. [1, A, p. 1], [3, K, pp. 28,30,31]

137. U.S. Department of Health and Human Services. 1985. Guidelines for Construction and Equipment of Hospitals and Medical Facilities. DHHS Publication No. (HRS-M-HF) 84-1. Washington, DC: U.S. Government Printing Office. [5, B, p. 48]

138. U.S. Department of Labor. 1983. Profile of Laboratories with the Potential for Exposure to Toxic Substances. Occupational Safety and Health Administration. Washington, DC: U.S. Government Printing Office. [2, B, p. 8]

139. U.S. Department of Labor and the U.S. Department of Health and Human Services. 1987. Department of Labor Joint Advisory Notice: Protection Against Occupational Exposure to Hepatitis B Virus and Human Immunodeficiency Virus. Federal Register 52(210):41818-41824. [2, C, p. 11]

140. Vesley D and Lauer J. 1986. Decontamination, sterilization, disinfection and antisepsis in the microbiological laboratory. Pp. 188-189 in Laboratory Safety: Principles and Practices,

BM Miller, DHM Groschel, JH Richardson, D Vesley, JR Songer, RD Housewright, and WE Barkley, eds. Washington, DC: American Society for Microbiology. [4, E, pp. 39,42]

141. Wedum AG. 1973. Proceedings of the National Cancer Institute Symposium on Centrifuge Biohazards. Cancer Research Safety Monograph Series. Volume 1. Bethesda, MD: National Cancer Institute. [3, I, p. 24]

142. Wedum AG, Barkley WE, and Hellman, A. 1972. Handling of infectious agents. Journal of the American Veterinary Medical Association 161:1557-1567. [2, C, pp. 11,12]

143. Weiss SH, Goedert JJ, Gartner S, Popovic M, Waters D, Markham P, Veronese FDM, Gail MH, Barkley WE, Gibbons J, Gill FA, Leuther M, Shaw GM, Gallo RC, and Blattner WA. 1988. Risk of human immunodeficiency virus (HIV-1) infection among laboratory workers. Science 239:68-71. [2, C, p. 12], [3, B, p. 13]

144. Welch LS. 1986. Decisionmaking about reproductive hazards. Seminars in Occupational Medicine 1:97-106. [5, D, p. 56]

145. West DL and Chatigny MA. 1986. Design of microbiological and biomedial research facilities. Pp. 124-137 in Laboratory Safety: Principles and Practices, BM Miller, DHM Groschel, JH Richardson, D Vesley, JR Songer, RD Housewright, and WE Barkley, eds. Washington, DC: American Society for Microbiology. [5, B, p. 48]

146. Wirtz RA. 1980. Occupational allergies to arthropods—documentation and prevention. Bulletin of the Entomological Society of America 26:356-360. [3, C, p. 15]

147. World Health Organization. 1983. Handling, transfer and shipment of specimens. Pp. 15-18 in Laboratory Biosafety Manual. Geneva: World Health Organization. [3, G, pp. 20,22]

148. World Health Organization. 1983. Safe shipment of specimens and infectious substances. Pp. 52-55 in Laboratory Biosafety Manual. Geneva: World Health Organization. [3, G, pp. 20,22]

149. World Health Organization. 1983. Laboratory Biosafety Manual. Geneva: World Health Organization. [Preface, p. vii], [5, C, p. 52]

150. Wolf WW. 1984. Controlling respiratory hazards in insectaries. Pp. 64-69 in Advances and Challenges in Insect Rearing, EG King and NC Leppla, eds. U.S. Department of Agriculture Technical Bulletin. [3, C, p. 15]

151. Wolf WW. 1985. Recognition and prevention of health hazards associated with insect rearing. Pp. 157-165 in Volume 1: Handbook of Insect Rearing, P Singh and RF Moore, eds. The Netherlands: Elsevier Scientific Publishers B.V. [3, C, pp. 15]

APPENDIXES

Appendix A

Biosafety in Microbiological and Biomedical Laboratories

CONTENTS

Reprinted from J.H. Richardson and W.E. Barkley, *Biosafety in Microbiological and Biomedical Laboratories,* 1st ed., U.S. Public Health Service, HHS Publication No. (CDC) 84-8395, U.S. Government Printing Office, Washington, D.C., March 1984. Figures, tables, appendixes, and page numbers in this reprinted version have been renumbered to avoid confusion.

SECTION I. INTRODUCTION

Microbiology laboratories are special, often unique, work environments that may pose special infectious disease risks to persons in or near them. Personnel have contracted infections in the laboratory throughout the history of microbiology. Published reports around the turn of the century described laboratory-associated cases of typhoid, cholera, glanders, brucellosis, and tetanus (123). In 1941, Meyer and Eddie (75) published a survey of 74 laboratory-associated brucellosis infections that had occurred in the United States and concluded that the "handling of cultures or specimens or the inhalation of dust containing *Brucella* organisms is eminently dangerous to laboratory workers." A number of cases were attributed to carelessness or poor technique in the handling of infectious materials.

In 1949, Sulkin and Pike (113) published the first in a series of surveys of laboratory-associated infections summarizing 222 viral infections—21 of which were fatal. In at least a third of the cases the probable source of infection was considered to be associated with the handling of infected animals and tissues. Known accidents were recorded in 27 (12%) of the reported cases.

In 1951, Sulkin and Pike (114) published the second of a series of summaries of laboratory-associated infections based on a questionnaire sent to 5,000 laboratories. Only one-third of the 1,342 cases cited had been reported in the literature. Brucellosis outnumbered all other reported laboratory-acquired infections and together with tuberculosis, tularemia, typhoid, and streptococcal infection accounted for 72% of all bacteria infections and for 31% of infections caused by all agents. The overall case fatality rate was 3%. Only 16% of all infections reported were associated with a documented accident. The majority of these were related to mouth pipetting and the use of needle and syringe.

This survey was updated in 1965 (93), adding 641 new or previously unreported cases, and again in 1976 (90), summarizing a cumulative total of 3,921 cases. Brucellosis, typhoid, tularemia, tuberculosis, hepatitis, and Venezuelan equine encephalitis were the most commonly reported. Fewer than 20% of all cases were associated with a known accident. Exposure to infectious aerosols was considered to be a plausible but unconfirmed source of infection for the more than 80% of the reported cases in which the infected person had "worked with the agent."

In 1967, Hanson et al. (53) reported 428 overt laboratory-associated infections with arboviruses. In some instances the ability of a given arbovirus to produce human disease was first confirmed as the result of unintentional infection of laboratory personnel. Exposure to infectious aerosols was considered the most common source of infection.

In 1974, Skinhoj (104) published the results of a survey which showed that personnel in Danish clinical chemistry laboratories had a reported incidence of hepatitis (2.3 cases per year per 1,000 employees) seven times higher than that of the general population. Similarly, a 1976 survey by Harrington and Shannon (55) indicated that medical laboratory workers in England had "a five times increased risk of acquiring tuberculosis compared with the general population." Hepatitis and shigellosis were also shown to be continuing occupational risks and together with tuberculosis were the three most commonly reported occupation-associated infections in Britain.

Although these reports suggest that laboratory personnel are at increased risk of being infected by the agents they handle, actual rates of infection are typically not available. However, the studies of Harrington and Shannon (55) and of Skinhoj (104) indicate that laboratory personnel have higher rates of tuberculosis, shigellosis, and hepatitis than the general population.

In contrast to the documented occurrence of laboratory-acquired infections in laboratory personnel, laboratories working with infectious agents have not been shown to represent a threat to the community. For example, although 109 laboratory-associated infections were recorded at the Center for Disease Control in 1947-1973 (97), no secondary cases were reported in family members or community contacts. The National Animal Disease Center has reported a similar experience (115), with no secondary cases occurring in laboratory and nonlaboratory contacts of 18 laboratory-associated cases occurring in 1960-1975. A secondary case of Marburg disease in the wife of a primary case was presumed to have been transmitted sexually two months after his dismissal from the hospital (70). Three secondary cases of smallpox were reported in two laboratory-associated outbreaks in England in 1973 (96) and 1978 (130). There were earlier reports of six cases of Q fever in employees of a commercial laundry which handled linens and uniforms from a laboratory where work with the agent was conducted (84), and two cases of Q fever in household contacts of a rickettsiologist (5). These cases are representative of the sporadic nature and infrequent association of community infections with laboratories working with infectious agents.

In his 1979 review (92), Pike concluded, "the knowledge, the techniques, and the equipment to prevent most laboratory infections are available." No single code of practice, standards, guidelines, or other publication, however, provides detailed descriptions of techniques, equipment, and other considerations or recommendations for the broad scope of laboratory activities conducted in the United States with a variety of indigenous and exotic infectious agents. The booklet *Classification of Etiologic Agents on the Basis of Hazard* (15) has, since 1969, served as a general reference for some laboratory activities utilizing infectious agents. That booklet and the concept of categorizing infectious agents and laboratory activities into four classes or levels served as a basic format for *Biosafety in Microbiological and Biomedical Laboratories*. This publication will provide specific descriptions of combinations of microbiological practices, laboratory facilities, and safety equipment and recommendations for use in four categories or biosafety levels of laboratory operation with selected infectious agents of man.

The descriptions of biosafety levels 1-4 parallel those of P1-4 in the NIH *Guidelines for Research Involving Recombinant DNA Molecules* (43) and are consistent with the general criteria used in assigning agents to Classes 1-4 in *Classification of Etiologic Agents on the Basis of Hazard* (15). Four biosafety levels are also described for infectious disease activities utilizing small laboratory animals. Recommendations for biosafety levels for specific agents are made on the basis of the potential hazard of the agent and of the laboratory function or activity.

SECTION II. PRINCIPLES OF BIOSAFETY

The term "containment" is used in describing safe methods for managing infectious agents in the laboratory environment where they are being handled or maintained. Primary containment, the protection of personnel and the immediate laboratory environment from exposure to infectious agents, is provided by good microbiological technique and the use of appropriate safety equipment. The use of vaccines may provide an increased level of personal protection. Secondary containment, the protection of the environment external to the laboratory from exposure to infectious materials, is provided by a combination of facility design and operational practices. The purpose of containment is to reduce exposure of laboratory workers and other persons to, and to prevent escape into the outside environment of, potentially hazardous agents. The three elements of containment include laboratory practice and technique, safety equipment, and facility design.

Laboratory practice and technique. The most important element of containment is strict adherence to standard microbiological practices and techniques. Persons working with infectious agents or infected materials must be aware of potential hazards and must be trained and proficient in the practices and techniques required for safely handling such material. The director or person in charge of the laboratory is responsible for providing or arranging for appropriate training of personnel.

When standard laboratory practices are not sufficient to control the hazard associated with a particular agent or laboratory procedure, additional measures may be needed. The laboratory director is responsible for selecting additional safety practices, which must be in keeping with the hazard associated with the agent or procedure.

Each laboratory should develop or adopt a biosafety or operations manual which identifies the hazards that will or may be encountered and which specifies practices and procedures designed to minimize or eliminate risks. Personnel should be advised of special hazards and should be required to read and to follow the required practices and procedures. A scientist trained and knowledgeable in appropriate laboratory techniques, safety procedures, and hazards associated with handling infectious agents must direct laboratory activities.

Laboratory personnel, safety practices, and techniques must be supplemented by appropriate facility design and engineering features, safety equipment, and management practices.

Safety equipment (primary barriers). Safety equipment includes biological safety cabinets and a variety of enclosed containers. The biological safety cabinet is the principal device used to provide containment of infectious aerosols generated by many microbiological procedures. Three types of biological safety cabinets (Class I, II, III) used in microbiological laboratories are described and illustrated in Appendix A.1. Open-fronted Class I and Class II biological safety cabinets are partial containment cabinets which offer significant levels of protection to laboratory personnel and to the environment when used with good microbiological techniques. The gastight Class III biological safety cabinet provides the highest attainable level of protection to personnel and the environment.

An example of an enclosed container is the safety centrifuge cup, which is designed to prevent aerosols from being released during centrifugation.

Safety equipment also includes items for personal protection such as gloves, coats, gowns, shoe covers, boots, respirators, face shields, and safety glasses. These personal protective devices are often used in combination with biological safety cabinets and other devices which contain the agents, animals, or materials being worked with. In some situations in which it is impractical to work in biological safety cabinets, personal protective devices may form the primary barrier between personnel and the infectious materials. Examples of such activities include certain animal studies, animal necropsy, production activities, and activities relating to maintenance, service, or support of the laboratory facility.

Facility design (secondary barriers). The design of the facility is important in providing a barrier to protect persons working in the facility but outside the laboratory and those in the community from infectious agents which may be accidentally released from the laboratory. Laboratory management is responsible for providing facilities commensurate with the laboratory's function. Three facility designs are described below, in ascending order by level of containment.

1. *The basic laboratory.* This laboratory provides general space in which work is done with viable agents which are not associated with disease in healthy adults. Basic laboratories include those facilities described in the following pages as Biosafety Levels 1 and 2 facilities.

This laboratory is also appropriate for work with infectious agents or potentially infectious materials when the hazard levels are low and laboratory personnel can be adequately protected by standard laboratory practice. While work is commonly conducted on the open bench, certain operations are confined to biological safety cabinets. Conventional laboratory designs are adequate. Areas known to be sources of general contamination, such as animal rooms and waste staging areas, should not be adjacent to patient care activities. Public areas and general offices to which nonlaboratory staff require frequent access should be separated from spaces which primarily support laboratory functions.

2. *The containment laboratory.* This laboratory has special engineering features which make it possible for laboratory workers to handle hazardous materials without endangering themselves, the community, or the environment. The containment laboratory is described in the following pages as a Biosafety Level 3 facility. The unique features which distinguish this laboratory from the basic laboratory are the provisions for access control and a specialized ventilation system. The containment laboratory may be an entire building or a single module or complex of modules within a building. In all cases, the laboratory is separated by a controlled access zone from areas open to the public.

3. *The maximum containment laboratory.* This laboratory has special engineering and containment features that allow activities involving infectious agents that are extremely hazardous to the laboratory worker or that may cause serious epidemic disease to be conducted safely. The maximum containment laboratory is described on the following pages as a Biosafety Level 4 facility. Although the maximum containment laboratory is generally a separate building, it can be constructed as an isolated area within a building. The laboratory's distinguishing characteristic is that it has secondary barriers to prevent hazardous materials from escaping into the environment. Such barriers include sealed openings into the laboratory, airlocks or liquid disinfectant barriers, a clothing-change and shower room contiguous to the laboratory ventilation system, and a treatment system to decontaminate exhaust air.

Biosafety levels. Four biosafety levels are described which consist of combinations of laboratory practices and techniques, safety equipment, and laboratory facilities appropriate for the operations performed and the hazard posed by the infectious agents and for the laboratory function or activity.

Biosafety Level 1. Biosafety level 1 practices, safety equipment, and facilities are appropriate for undergraduate and secondary educational training and teaching laboratories and for other facilities in which work is done with defined and characterized strains of viable microorganisms not known to cause disease in healthy adult humans. *Bacillus subtilis, Naegleria gruberi,* and infectious canine hepatitis virus are representative of those microorganisms meeting these criteria. Many agents not ordinarily associated with disease processes in humans are, however, opportunistic pathogens and may cause infection in the young, the aged, and immunodeficient or immunosuppressed individuals. Vaccine strains which have undergone multiple in vivo passages should not be considered avirulent simply because they are vaccine strains.

Biosafety Level 2. Biosafety Level 2 practices, equipment, and facilities are applicable to clinical, diagnostic, teaching, and other facilities in which work is done with the broad spectrum of indigenous moderate-risk agents present in the community and associated with human disease of varying severity. With good microbiological techniques, these agents can be used safely in activities conducted on the open bench, provided the potential for producing aerosols is low. Hepatitis B virus, the salmonellae, and *Toxoplasma* spp. are representative of microorganisms assigned to this containment level. Primary hazards to personnel working with these agents may include accidental autoinoculation, ingestion, and skin or mucous membrane exposure to infectious materials. Procedures with high aerosol potential that may increase the risk of exposure of personnel must be conducted in primary containment equipment or devices.

Biosafety Level 3. Biosafety Level 3 practices, safety equipment, and facilities are applicable to clinical, diagnostic, teaching, research, or production facilities in which work is done with indigenous or exotic agents where the potential for infection by aerosols is real and the disease may have serious or lethal consequences. Autoinoculation and ingestion also represent primary hazards to personnel working with these agents. Examples of such agents for which Biosafety Level 3 safeguards are generally recommended include *Mycobacterium tuberculosis,* St. Louis encephalitis virus, and *Coxiella burnetii.*

Biosafety Level 4. Biosafety Level 4 practices, safety equipment, and facilities are applicable to work with dangerous and exotic agents which pose a high individual risk of life-threatening disease. All manipulations of potentially infectious diagnostic materials, isolates, and naturally or experimentally infected animals pose a high risk of exposure and infection to laboratory personnel. Lassa fever virus is representative of the microorganisms assigned to Level 4.

Animal biosafety levels. Four biosafety levels are also described for activities involving infectious disease activities with experimental mammals. These four combinations of practices, safety equipment, and facilities are designated *Animal Biosafety Levels* 1, 2, 3, and 4 and provide increasing levels of protection to personnel and the environment.

The laboratory director is directly and primarily responsible for the safe operation of the laboratory. His/her knowledge and judgment are critical in assessing risks and appropriately applying these recommendations. The recommended biosafety level represents those conditions under which the agent can ordinarily be safely handled. Special characteristics of the agents used, the training and experience of personnel, and the nature or function of the laboratory may further influence the director in applying these recommendations.

Work with known agents should be conducted at the biosafety level recommended in Section V unless specific information is available to suggest that virulence, pathogenicity, antibiotic resistance patterns, and other factors are significantly altered to require more stringent or allow less stringent practices to be used.

Clinical laboratories, and especially those in health care facilities, receive clinical specimens with requests for a variety of diagnostic and clinical support services. Typically, clinical laboratories receive specimens without pertinent information such as patient history or clinical findings which may be suggestive of an infectious etiology. Furthermore, such specimens are often submitted with a broad request for microbiological examination for multiple agents (e.g., sputum samples submitted for "routine," acid-fast, and fungal cultures).

It is the responsibility of the laboratory director to establish standard procedures in the laboratory which realistically address the issue of the infective hazard of clinical specimens. Except in extraordinary circumstances (e.g., suspected hemorrhagic fever) the initial processing of clinical specimens and identification of isolates can be and are safely conducted using a combination of practices, facilities, and safety equipment described as Biosafety Level 2. Biological safety cabinets (Class I or II) should be used for the initial processing of clinical specimens when the nature of the test requested or other information is suggestive that an agent readily transmissible by infectious aerosols is likely to be present. Class II biological safety cabinets are also used to protect the integrity of the specimens or cultures by preventing contamination from the laboratory environment.

Segregating clinical laboratory functions and limiting or restricting access to laboratory areas are the responsibility of the laboratory director.

Importation and interstate shipment of certain biomedical materials. The importation of etiologic agents and vectors of human diseases is subject to the requirements of the Public Health Service Foreign Quarantine regulations. Companion regulations of the Public Health Service and the Department of Transportation specify packaging, labeling, and shipping requirements for etiologic agents and diagnostic specimens shipped in interstate commerce (see Appendix A.4).

The U.S. Department of Agriculture regulates the importation and interstate shipment of animal pathogens and prohibits the importation, possession, or use of certain exotic animal disease agents which pose a serious disease threat to domestic livestock and poultry (see Appendix A.5).

SECTION III. LABORATORY BIOSAFETY LEVEL CRITERIA

The essential elements of the four biosafety levels for activities involving infectious microorganisms and laboratory animals are summarized in Tables A.1 and A.2. The levels are designated in ascending order by degree of protection provided to personnel, the environment, and the community.

Biosafety Level 1. Biosafety Level 1 is suitable for work involving agents of no known or minimal potential hazard to laboratory personnel and the environment. The laboratory is not separated from the general traffic patterns in the building. Work is generally conducted on open bench tops. Special containment equipment is not required or generally used. Laboratory personnel have specific training in the procedures conducted in the laboratory and are supervised by a scientist with general training in microbiology or a related science.

The following standard and special practices, safety equipment, and facilities apply to agents assigned to Biosafety Level 1.

TABLE A.1 Summary of Recommended Biosafety Levels for Infectious Agents

Biosafety level	Practices and techniques	Safety equipment	Facilities
1	Standard microbiological practices	None: primary containment provided by adherence to standard laboratory practices during open bench operations	Basic
2	Level 1 practices plus: laboratory coats; decontamination of all infectious wastes; limited access; protective gloves and biohazard warning signs as indicated	Partial containment equipment (i.e., Class I or II Biological Safety Cabinets) used to conduct mechanical and manipulative procedures that have high aerosol potential which may increase the risk of exposure to personnel	Basic
3	Level 2 practices plus: special laboratory clothing; controlled access	Partial containment equipment used for all manipulations of infectious material	Containment
4	Level 3 practices plus: entrance through change room where street clothing is removed and laboratory clothing is put on; shower on exit; all wastes are decontaminated on exit from the facility	Maximum containment equipment (i.e., Class III biological safety cabinet or partial containment equipment in combination with full-body, air-supplied, positive-pressure personnel suit) used for all procedures and activities	Maximum containment

TABLE A.2 Summary of Recommended Biosafety Levels for Activities in Which Experimentally or Naturally Infected Vertebrate Animals Are Used

Biosafety level	Practices and techniques	Safety equipment	Facilities
1	Standard animal care and management practices	None	Basic
2	Laboratory coats; decontamination of all infectious wastes and of animal cages prior to washing; limited access; protective gloves and hazard warning signs as indicated	Partial containment equipment and/or personal protective devices used for activities and manipulations of agents or infected animals that produce aerosols	Basic
3	Level 2 practices plus: special laboratory clothing; controlled access	Partial containment equipment and/or personal protective devices used for all activities and manipulations of agents or infected animals	Containment
4	Level 3 practices plus: entrance through clothes change room where street clothing is removed and laboratory clothing is put on shower on exit; all wastes are decontaminated before removal from the facility	Maximum containment equipment (i.e., Class III biological safety cabinet or partial containment equipment in combination with full-body, air supplied, positive-pressure personnel suit) used for all procedures and activities	Maximum containment

A. *Standard microbiological practices*

1. Access to the laboratory is limited or restricted at the discretion of the laboratory director when experiments are in progress.
2. Work surfaces are decontaminated once a day and after any spill of viable material.
3. All contaminated liquid or solid wastes are decontaminated before disposal.
4. Mechanical pipetting devices are used; mouth pipetting is prohibited.
5. Eating, drinking, smoking, and applying cosmetics are not permitted in the work area. Food may be stored in cabinets or refrigerators designated and used for this purpose only. Food storage cabinets or refrigerators should be located outside of the work area.
6. Persons wash their hands after they handle viable materials and animals and before leaving the laboratory.
7. All procedures are performed carefully to minimize the creation of aerosols.
8. It is recommended that laboratory coats, gowns, or uniforms be worn to prevent contamination or soiling of street clothes.

B. *Special practices*

1. Contaminated materials that are to be decontaminated at a site away from the laboratory are placed in a durable leakproof container which is closed before being removed from the laboratory.
2. An insect and rodent control program is in effect.

C. *Containment equipment*

Special containment equipment is generally not required for manipulations of agents assigned to Biosafety Level 1.

D. *Laboratory facilities*

1. The laboratory is designed so that it can be easily cleaned.
2. Bench tops are impervious to water and resistant to acids, alkalis, organic solvents, and moderate heat.
3. Laboratory furniture is sturdy. Spaces between benches, cabinets, and equipment are accessible for cleaning.
4. Each laboratory contains a sink for handwashing.
5. If the laboratory has windows that open, they are fitted with fly screens.

Biosafety Level 2. Biosafety Level 2 is similar to Level 1 and is suitable for work involving agents of moderate potential hazard to personnel and the environment. It differs in that (1) laboratory personnel have specific training in handling pathogenic agents and are directed by competent scientists, (2) access to the laboratory is limited when work is being conducted, and (3) certain procedures in which infectious aerosols are created are conducted in biological safety cabinets or other physical containment equipment.

The following standard and special practices, safety equipment, and facilities apply to agents assigned to Biosafety Level 2.

A. *Standard microbiological practices*

1. Access to the laboratory is limited or restricted by the laboratory director when work with infectious agents is in progress.

2. Work surfaces are decontaminated at least once a day and after any spill of viable material.

3. All infectious liquid or solid wastes are decontaminated before disposal.

4. Mechanical pipetting devices are used; mouth pipetting is prohibited.

5. Eating, drinking, smoking, and applying cosmetics are not permitted in the work area. Food may be stored in cabinets or refrigerators designated and used for this purpose only. Food storage cabinets or refrigerators should be located outside of the work area.

6. Persons wash their hands after handling infectious materials and animals and when they leave the laboratory.

7. All procedures are performed carefully to minimize the creation of aerosols.

B. *Special practices*

1. Contaminated materials that are to be decontaminated at a site away from the laboratory are placed in a durable leakproof container which is closed before being removed from the laboratory.

2. The laboratory director limits access to the laboratory. In general, persons who are at increased risk of acquiring infection or for whom infection may be unusually hazardous are not allowed in the laboratory or animal rooms. The director has the final responsibility for assessing each circumstance and determining who may enter or work in the laboratory.

3. The laboratory director establishes policies and procedures whereby only persons who have been advised of the potential hazard and meet any specific entry requirements (e.g., immunization) enter the laboratory or animal rooms.

4. When the infectious agent(s) in use in the laboratory require special provisions for entry (e.g., vaccination), a hazard warning sign, incorporating the universal biohazard symbol, is posted on the access door to the laboratory work area. The hazard warning sign identifies the infectious agent, lists the name and telephone number of the laboratory director or other responsible person(s), and indicates the special requirement(s) for entering the laboratory.

5. An insect and rodent control program is in effect.

6. Laboratory coats, gowns, smocks, or uniforms are worn while in the laboratory. Before leaving the laboratory for nonlaboratory areas (e.g., cafeteria, library, administrative offices), this protective clothing is removed and left in the laboratory or covered with a clean coat not used in the laboratory.

7. Animals not involved in the work being performed are not permitted in the laboratory.

8. Special care is taken to avoid skin contamination with infectious materials; gloves should be worn when handling infected animals and when skin contact with infectious materials is unavoidable.

9. All wastes from laboratories and animal rooms are appropriately decontaminated before disposal.

10. Hypodermic needles and syringes are used only for parenteral injection and aspiration of fluids from laboratory animals and diaphragm bottles. Only needle-locking syringes or disposable syringe-needle units (i.e., needle is integral to the syringe) are used for the injection or aspiration of infectious fluids. Extreme caution should be used when handling needles and syringes to avoid autoinoculation and the generation of aerosols during use and disposal. Needles should not be bent, sheared, replaced in the sheath or guard, or removed from the syringe following use. The needle and syringe should be promptly placed in a puncture-resistant container and decontami-

nated, preferably by autoclaving, before discard or reuse.

11. Spills and accidents which result in overt exposures to infectious materials are immediately reported to the laboratory director. Medical evaluation, surveillance, and treatment are provided as appropriate, and written records are maintained.

12. When appropriate, considering the agent(s) handled, base-line serum samples for laboratory and other at-risk personnel are collected periodically, depending on the agents handled or the function of the facility.

13. A biosafety manual is prepared or adopted. Personnel are advised of special hazards and are required to read instructions on practices and procedures and to follow them.

C. *Containment equipment*

Biological safety cabinets (Class I or II) (see Appendix A.1) or other appropriate personal protective or physical containment devices are used whenever:

1. Procedures with a high potential for creating infectious aerosols are conducted (82). These may include centrifuging, grinding, blending, vigorous shaking or mixing, sonic disruption, opening containers of infectious materials whose internal pressures may be different from ambient pressures, inoculating animals intranasally, and harvesting infected tissues from animals or eggs.

2. High concentrations or large volumes of infectious agents are used. Such materials may be centrifuged in the open laboratory if sealed heads or centrifuge safety cups are used and if they are opened only in a biological safety cabinet.

D. *Laboratory facilities*

1. The laboratory is designed so that it can be easily cleaned.

2. Bench tops are impervious to water and resistant to acids, alkalis, organic solvents, and moderate heat.

3. Laboratory furniture is sturdy, and spaces between benches, cabinets, and equipment are accessible for cleaning.

4. Each laboratory contains a sink for hand-washing.

5. If the laboratory has windows that open, they are fitted with fly screens.

6. An autoclave for decontaminating infectious laboratory wastes is available.

Biosafety Level 3. Biosafety Level 3 is applicable to clinical, diagnostic, teaching, research, or production facilities in which work is done with indigenous or exotic agents which may cause serious or potentially lethal disease as a result of exposure by the inhalation route. Laboratory personnel have specific training in handling pathogenic and potentially lethal agents and are supervised by competent scientists who are experienced in working with these agents. All procedures involving the manipulation of infectious material are conducted within biological safety cabinets or other physical containment devices or by personnel wearing appropriate personal protective clothing and devices. The laboratory has special engineering and design features. It is recognized, however, that many existing facilities may not have all the facility safeguards recommended for Biosafety Level 3 (e.g., access zone, sealed penetrations, directional airflow, etc.). In these circumstances, acceptable safety may be achieved for routine or repetitive operations (e.g., diagnostic procedures involving the propagation of an agent for identification, typing, and susceptibility testing) in laboratories where facility features satisfy Biosafety Level 2 recommendations, provided the recommended "Standard Microbiological Practices," "Special Practices," and "Containment Equipment" for Biosafety Level 3 are rigorously followed. The decision to implement this modification of Biosafety Level 3 recommendations should be made only by the laboratory director.

The following standard and special safety practices, equipment, and facilities apply to agents assigned to Biosafety Level 3.

A. *Standard microbiological practices*

1. Work surfaces are decontaminated at least once a day and after any spill of viable material.

2. All infectious liquid or solid wastes are decontaminated before disposal.

3. Mechanical pipetting devices are used; mouth pipetting is prohibited.

4. Eating, drinking, smoking, storing food, and applying cosmetics are not permitted in the work area.

5. Persons wash their hands after handling infectious materials and animals and when they leave the laboratory.

6. All procedures are performed carefully to minimize the creation of aerosols.

B. *Special practices*

1. Laboratory doors are kept closed when experiments are in progress.

2. Contaminated materials that are to be decontaminated at a site away from the laboratory are placed in a durable leakproof container which is closed before being removed from the laboratory.

3. The laboratory director controls access to the laboratory and restricts access to persons whose presence is required for program or support purposes. Persons who are at increased risk of acquiring infection or for whom infection may be unusually hazardous are not allowed in the laboratory or animal rooms. The director has the final responsibility for assessing each circumstance and determining who may enter or work in the laboratory.

4. The laboratory director establishes policies and procedures whereby only persons who have been advised of the potential biohazard, who meet any specific entry requirements (e.g., immunization), and who comply with all entry and exit procedures enter the laboratory or animal rooms.

5. When infectious materials or infected animals are present in the laboratory or containment module, a hazard warning sign, incorporating the universal biohazard symbol, is posted on all laboratory and animal room access doors. The hazard warning sign identifies the agent, lists the name and telephone number of the laboratory director or other responsible person(s), and indicates any special requirements for entering the laboratory, such as the need for immunizations, respirators, or other personal protective measures.

6. All activities involving infectious materials are conducted in biological safety cabinets or other physical containment devices within the containment module. No work in open vessels is conducted on the open bench.

7. The work surfaces of biological safety cabinets and other containment equipment are decontaminated when work with infectious materials is finished. Plastic-backed paper toweling used on nonperforated work surfaces within biological safety cabinets facilitates cleanup.

8. An insect and rodent control program is in effect.

9. Laboratory clothing that protects street clothing (e.g., solid front or wrap-around gowns, scrub suits, coveralls) is worn in the laboratory. Laboratory clothing is not worn outside the laboratory, and it is decontaminated before being laundered.

10. Special care is taken to avoid skin contamination with infectious materials; gloves should be worn when handling infected animals and when skin contact with infectious materials is unavoidable.

11. Molded surgical masks or respirators are worn in rooms containing infected animals.

12. Animals and plants not related to the work being conducted are not permitted in the laboratory.

13. All wastes from laboratories and animal rooms are appropriately decontaminated before disposal.

14. Vacuum lines are protected with high efficiency particulate air (HEPA) filters and liquid disinfectant traps.

15. Hypodermic needles and syringes are used only for parenteral injection and aspiration of fluids from laboratory animals and diaphragm bottles. Only needle-locking syringes or disposable syringe-needle units (i.e., needle is integral to the syringe) are used for the injection or aspiration of infectious fluids. Extreme caution should be used when handling needles and syringes to avoid autoinoculation and the generation of aerosols during use and disposal. Needles should not be bent, sheared, replaced in the sheath or guard, or removed from the syringe following use. The needle and syringe should be promptly placed in a puncture-resistant container and decontaminated, preferably by autoclaving, before discard or reuse.

16. Spills and accidents which result in overt or potential exposures to infectious materials are

immediately reported to the laboratory director. Appropriate medical evaluation, surveillance, and treatment are provided and written records are maintained.

17. Base-line serum samples for all laboratory and other at-risk personnel should be collected and stored. Additional serum specimens may be collected periodically, depending on the agents handled or the function of the laboratory.

18. A biosafety manual is prepared or adopted. Personnel are advised of special hazards and are required to read instructions on practices and procedures and to follow them.

C. *Containment equipment*

Biological safety cabinets (Class I, II, or III) (see Appendix A.1) or other appropriate combinations of personal protective or physical containment devices (e.g., special protective clothing, masks, gloves, respirators, centrifuge safety cups, sealed centrifuge rotors, and containment caging for animals) are used for all activities with infectious materials which pose a threat of aerosol exposure. These include: manipulation of cultures and of those clinical or environmental materials which may be a source of infectious aerosols; the aerosol challenge of experimental animals; harvesting of tissues or fluids from infected animals and embryonated eggs; and necropsy of infected animals.

D. *Laboratory facilities*

1. The laboratory is separated from areas which are open to unrestricted traffic flow within the building. Passage through two sets of doors is the basic requirement for entry into the laboratory from access corridors or other contiguous areas. Physical separation of the high containment laboratory from access corridors or other laboratories or activities may also be provided by a double-doored clothes change room (showers may be included), airlock, or other access facility which requires passage through two sets of doors before entering the laboratory.

2. The interior surfaces of walls, floors, and ceilings are water resistant so that they can be easily cleaned. Penetrations in these surfaces are sealed or capable of being sealed to facilitate decontaminating the area.

3. Bench tops are impervious to water and resistant to acids, alkalis, organic solvents, and moderate heat.

4. Laboratory furniture is sturdy, and spaces between benches, cabinets, and equipment are accessible for cleaning.

5. Each laboratory contains a sink for handwashing. The sink is foot, elbow, or automatically operated and is located near the laboratory exit door.

6. Windows in the laboratory are closed and sealed.

7. Access doors to the laboratory or containment module are self-closing.

8. An autoclave for decontaminating laboratory wastes is available, preferably within the laboratory.

9. A ducted exhaust air ventilation system is provided. This system creates directional airflow that draws air into the laboratory through the entry area. The exhaust air is not recirculated to any other area of the building, is discharged to the outside, and is dispersed away from occupied areas and air intakes. Personnel must verify that the direction of the airflow (into the laboratory) is proper. The exhaust air from the laboratory room can be discharged to the outside without being filtered or otherwise treated.

10. The HEPA-filtered exhaust air from Class I or Class II biological safety cabinets is discharged directly to the outside or through the building exhaust system. Exhaust air from Class I or II biological safety cabinets may be recirculated within the laboratory if the cabinet is tested and certified at least every 12 months. If the HEPA-filtered exhaust air from Class I or II biological safety cabinets is to be discharged to the outside through the building exhaust air system, it is connected to this system in a manner (e.g., thimble unit connection [80]) that avoids any interference with the air balance of the cabinets or building exhaust system.

Biosafety Level 4. Biosafety Level 4 is required for work with dangerous and exotic agents which pose a high individual risk of life-threatening disease. Members of the laboratory staff have specific and thorough training in handling extremely hazardous infectious agents, and they understand the primary and secondary containment functions of the standard and special practices, the containment equipment, and the laboratory design characteristics. They are supervised by competent scientists who are trained and experienced in working with these agents. Access to the laboratory is strictly controlled by the laboratory director. The facility is either in a separate building or in a controlled area within a building, which is completely isolated from all other areas of the building. A specific facility operations manual is prepared or adopted.

Within work areas of the facility, all activities are confined to Class III biological safety cabinets or Class I or Class II biological safety cabinets used along with one-piece positive-pressure personnel suits ventilated by a life support system. The maximum containment laboratory has special engineering and design features to prevent microorganisms from being disseminated into the environment.

The following standard and special safety practices, equipment, and facilities apply to agents assigned to Biosafety Level 4.

A. *Standard microbiological practices*

1. Work surfaces are decontaminated at least once a day and immediately after any spill of viable material.

2. Only mechanical pipetting devices are used.

3. Eating, drinking, smoking, storing food, and applying cosmetics are not permitted in the laboratory.

4. All procedures are performed carefully to minimize the creation of aerosols.

B. *Special practices*

1. Biological materials to be removed from the Class III cabinet or from the maximum containment laboratory in a viable or intact state are transferred to a nonbreakable, sealed primary container and then enclosed in a nonbreakable, sealed secondary container which is removed from the facility through a disinfectant dunk tank, fumigation chamber, or an airlock designed for this purpose.

2. No materials, except for biological materials that are to remain in a viable or intact state, are removed from the maximum containment laboratory unless they have been autoclaved or decontaminated before they leave the facility. Equipment or material which might be damaged by high temperatures or steam is decontaminated by gaseous or vapor methods in an airlock or chamber designed for this purpose.

3. Only persons whose presence in the facility or individual laboratory rooms is required for program or support purposes are authorized to enter. Persons who may be at increased risk of acquiring infection or for whom infection may be unusually hazardous are not allowed in the laboratory or animal rooms. The supervisor has the final responsibility for assessing each circumstance and determining who may enter or work in the laboratory. Access to the facility is limited by means of secure, locked doors; accessibility is managed by the laboratory director, biohazards control officer, or other person responsible for the physical security of the facility. Before entering, persons are advised of the potential biohazards and instructed as to appropriate safeguards for ensuring their safety. Authorized persons comply with the instructions and all other applicable entry and exit procedures. A logbook, signed by all personnel, indicates the date and time of each entry and exit. Practical and effective protocols for emergency situations are established.

4. Personnel enter and leave the facility only through the clothing change and shower rooms. Personnel shower each time they leave the facility. Personnel use the airlocks to enter or leave the laboratory only in an emergency.

5. Street clothing is removed in the outer clothing change room and kept there. Complete laboratory clothing, including undergarments, pants and shirts or jumpsuits, shoes, and gloves, is provided and used by all personnel entering the facility. Head covers are provided for personnel who do not wash their hair during the exit shower. When leaving the laboratory and before proceeding into the shower area, personnel remove their laboratory clothing and store it in a locker or hamper in the inner change room.

6. When infectious materials or infected animals are present in the laboratory or animal rooms, a hazard warning sign, incorporating the universal biohazard symbol, is posted on all access doors. The sign identifies the agent, lists the name and telephone number of the laboratory

director or other responsible person(s), and indicates any special requirements for entering the area (e.g., the need for immunizations or respirators).

7. Supplies and materials needed in the facility are brought in by way of the double-doored autoclave, fumigation chamber, or airlock which is appropriately decontaminated between each use. After securing the outer doors, personnel within the facility retrieve the materials by opening the interior doors of the autoclave, fumigation chamber, or airlock. These doors are secured after materials are brought into the facility.

8. An insect and rodent control program is in effect.

9. Materials (e.g., plants, animals, and clothing) not related to the experiment being conducted are not permitted in the facility.

10. Hypodermic needles and syringes are used only for parenteral injection and aspiration of fluids from laboratory animals and diaphragm bottles. Only needle-locking syringes or disposable syringe-needle units (i.e., needle is integral part of unit) are used for the injection or aspiration of infectious fluids. Needles should not be bent, sheared, replaced in the needle guard, or removed from the syringe following use. The needle and syringe should be placed in a puncture-resistant container and decontaminated, preferably by autoclaving, before discard or reuse. Whenever possible, cannulas are used instead of sharp needles (e.g., for gavage).

11. A system is set up for reporting laboratory accidents and exposures and employee absenteeism, and for the medical surveillance of potential laboratory-associated illnesses. Written records are prepared and maintained. An essential adjunct to such a reporting-surveillance system is the availability of a facility for the quarantine, isolation, and medical care of personnel with potential or known laboratory-associated illnesses.

C. *Containment equipment*

All procedures within the facility with agents assigned to Biosafety Level 4 are conducted in a Class III biological safety cabinet or in Class I or II biological safety cabinets used in conjunction with one-piece positive-pressure personnel suits ventilated by a life support system. Activities with viral agents (e.g., Rift Valley fever virus) that require Biosafety Level 4 secondary containment capabilities and for which highly effective vaccines are available and used can be conducted within Class I or Class II biological safety cabinets within the facility without the one-piece positive-pressure personnel suit being used, if (1) the facility has been decontaminated; (2) no work is being conducted in the facility with other agents assigned to Biosafety Level 4; and (3) all other standards and special practices are followed.

D. *Laboratory facilities*

1. The maximum containment facility consists of either a separate building or a clearly demarcated and isolated zone within a building. Outer and inner change rooms separated by a shower are provided for personnel entering and leaving the facility. A double-doored autoclave, fumigation chamber, or ventilated airlock is provided for passage of those materials, supplies, or equipment which are not brought into the facility through the change room.

2. Walls, floors, and ceilings of the facility are constructed to form a sealed internal shell which facilitates fumigation and is animal and insect proof. The internal surfaces of this shell are resistant to liquids and chemicals, thus facilitating cleaning and decontamination of the area. All penetrations in these structures and surfaces are sealed. Any drains in the floors contain traps filled with a chemical disinfectant of demonstrated efficacy against the target agent, and they are connected directly to the liquid waste decontamination system. Sewer and other ventilation lines contain HEPA filters.

3. Internal facility appurtenances, such as light fixtures, air ducts, and utility pipes, are arranged to minimize the horizontal surface area on which dust can settle.

4. Bench tops have seamless surfaces which are impervious to water and resistant to acids, alkalis, organic solvents, and moderate heat.

5. Laboratory furniture is of simple and sturdy construction, and spaces between benches, cabinets, and equipment are accessible for cleaning.

6. A foot-, elbow-, or automatically oper-

ated handwashing sink is provided near the door of each laboratory room in the facility.

7. If there is a central vacuum system, it does not serve areas outside the facility. In-line HEPA filters are placed as near as practicable to each use point or service cock. Filters are installed to permit in-place decontamination and replacement. Other liquid and gas services to the facility are protected by devices that prevent backflow.

8. If water fountains are provided, they are foot operated and are located in the facility corridors outside the laboratory. The water service to the fountain is not connected to the backflow-protected distribution system supplying water to the laboratory areas.

9. Access doors to the laboratory are self-closing and lockable.

10. Any windows are breakage resistant.

11. A double-doored autoclave is provided for decontaminating materials passing out of the facility. The autoclave door which opens to the area external to the facility is sealed to the outer wall and automatically controlled so that the outside door can only be opened after the autoclave "sterilization" cycle has been completed.

12. A pass-through dunk tank, fumigation chamber, or an equivalent decontamination method is provided so that materials and equipment that cannot be decontaminated in the autoclave can be safely removed from the facility.

13. Liquid effluents from laboratory sinks, biological safety cabinets, floors, and autoclave chambers are decontaminated by heat treatment before being released from the maximum containment facility. Liquid wastes from shower rooms and toilets may be decontaminated with chemical disinfectants or by heat in the liquid waste decontamination system. The procedure used for heat decontamination of liquid wastes is evaluated mechanically and biologically by using a recording thermometer and an indicator microorganism with a defined heat susceptibility pattern. If liquid wastes from the shower rooms are decontaminated with chemical disinfectants, the chemical used is of demonstrated efficacy against the target or indicator microorganisms.

14. An individual supply and exhaust air ventilation system is provided. The system maintains pressure differentials and directional airflow as required to assure flows inward from areas outside of the facility toward areas of highest potential risk within the facility. Manometers are used to sense pressure differentials between adjacent areas maintained at different pressure levels. If a system malfunctions, the manometers sound an alarm. The supply and exhaust airflow is interlocked to assure inward (or zero) airflow at all times.

15. The exhaust air from the facility is filtered through HEPA filters and discharged to the outside so that it is dispersed away from occupied buildings and air intakes. Within the facility, the filters are located as near the laboratories as practicable in order to reduce the length of potentially contaminated air ducts. The filter chambers are designed to allow in situ decontamination before filters are removed and to facilitate certification testing after they are replaced. Coarse filters and HEPA filters are provided to treat air supplied to the facility in order to increase the lifetime of the exhaust HEPA filters and to protect the supply air system should air pressures become unbalanced in the laboratory.

16. The treated exhaust air from Class I and II biological safety cabinets can be discharged into the laboratory room environment or to the outside through the facility air exhaust system. If exhaust air from Class I or II biological safety cabinets is discharged into the laboratory, the cabinets are tested and certified at 6-month intervals. *The treated exhaust air from Class III biological safety cabinets is discharged, without recirculation through two sets of HEPA filters in series, via the facility exhaust air system.* If the treated exhaust air from any of these cabinets is discharged to the outside through the facility exhaust air system, it is connected to this system in a manner (e.g., thimble unit connection [80]) that avoids any interference with the air balance of the cabinets or the facility exhaust air system.

17. A specially designed suit area may be provided in the facility. Personnel who enter this area wear a one-piece positive-pressure suit that is ventilated by a life support system. The life support system includes alarms and emergency backup breathing air tanks. Entry to this area is through an airlock fitted with airtight doors. A chemical shower is provided to decontaminate

the surface of the suit before the worker leaves the area. The exhaust air from the suit area is filtered by two sets of HEPA filters installed in series. A duplicate filtration unit, an exhaust fan, and an automatically starting emergency power source are provided. The air pressure within the suit area is lower than that of any adjacent area. Emergency lighting and communications systems are provided. All penetrations into the internal shell of the suit area are sealed. A double-doored autoclave is provided for decontaminating waste materials to be removed from the suit area.

SECTION IV. VERTEBRATE ANIMAL BIOSAFETY LEVEL CRITERIA

If experimental animals are used, institutional management must provide facilities and staff and establish practices which reasonably assure appropriate levels of environmental quality, safety, and care. Laboratory animal facilities are extensions of the laboratory and in some situations are integral to and inseparable from the laboratory. As a general principle, the Biosafety Level (facilities, practices, and operational requirements) recommended for working with infectious agents in vivo and in vitro are comparable.

These recommendations presuppose that laboratory animal facilities, operational practices, and quality of animal care meet applicable standards and regulations and that appropriate species have been selected for animal experiments (e.g., *Guide for the Care and Use of Laboratory Animals*, HEW Publication no. [NIH] 78-23, Rev. 1978, and *Laboratory Animal Welfare Regulations*, 9 CFR, Subchapter A, Parts 1, 2, and 3).

Ideally, facilities for laboratory animals used for studies of infectious or noninfectious disease should be physically separate from other activities such as animal production and quarantine, clinical laboratories, and especially from facilities that provide patient care. Animal facilities should be designed and constructed to facilitate cleaning and housekeeping. A "clean hall/dirty hall" layout is very useful in reducing cross contamination. Floor drains should be installed in animal facilities only on the basis of clearly defined needs. If floor drains are installed, the drain trap should always contain water.

These recommendations describe four combinations of practices, safety equipment, and facilities for experiments on animals infected with agents which are known or believed to produce infections in humans. These four combinations provide increasing levels of protection to personnel and to the environment and are recommended as minimal standards for activities involving infected laboratory animals. These four combinations, designated Animal Biosafety Levels 1-4, describe animal facilities and practices applicable to work on animals infected with agents assigned to corresponding Biosafety Levels 1-4.

Facility standards and practices for invertebrate vectors and hosts are not specifically addressed in standards written for commonly used laboratory animals. "Laboratory Safety for Arboviruses and Certain Other Viruses of Vertebrates" (112), prepared by the Subcommittee on Arbovirus Laboratory Safety of the American Committee on Arthropod-Borne Viruses, serves as a useful reference in the design and operation of facilities using arthropods.

Animal Biosafety Level 1

A. *Standard practices*

1. Doors to animal rooms open inward, are self-closing, and are kept closed when experimental animals are present.
2. Work surfaces are decontaminated after use or after any spill of viable materials.
3. Eating, drinking, smoking, and storing food for human use are not permitted in animal rooms.
4. Personnel wash their hands after handling cultures and animals and before leaving the animal room.
5. All procedures are carefully performed to minimize the creation of aerosols.
6. An insect and rodent control program is in effect.

B. *Special practices*

1. Bedding materials from animal cages are removed in such a manner as to minimize the creation of aerosols and disposed of in compliance with applicable institutional or local requirements.

2. Cages are washed manually or in a cage-washer. Temperature of final rinse water in a mechanical washer should be 180°F.

3. The wearing of laboratory coats, gowns, or uniforms in the animal room is recommended. It is further recommended that laboratory coats worn in the animal room not be worn in other areas.

C. *Containment equipment*

Special containment equipment is not required for animals infected with agents assigned to Biosafety Level 1.

D. *Animal facilities*

1. The animal facility is designed and constructed to facilitate cleaning and housekeeping.

2. A handwashing sink is available in the animal facility.

3. If the animal facility has windows that open, they are fitted with fly screens.

4. It is recommended, but not required, that the direction of airflow in the animal facility is inward and that exhaust air is discharged to the outside without being recirculated to other rooms.

Animal Biosafety Level 2

A. *Standard practices*

1. Doors to animal rooms open inward, are self-closing, and are kept closed when infected animals are present.

2. Work surfaces are decontaminated after use or spills of viable materials.

3. Eating, drinking, smoking, and storing of food for human use are not permitted in animal rooms.

4. Personnel wash their hands after handling cultures and animals and before leaving the animal room.

5. All procedures are carefully performed to minimize the creation of aerosols.

6. An insect and rodent control program is in effect.

B. *Special practices*

1. Cages are decontaminated, preferably by autoclaving, before they are cleaned and washed.

2. Surgical-type masks are worn by all personnel entering animal rooms housing nonhuman primates.

3. Laboratory coats, gowns, or uniforms are worn while in the animal room. This protective clothing is removed before leaving the animal facility.

4. The laboratory or animal facility director limits access to the animal room to personnel who have been advised of the potential hazard and who need to enter the room for program or service purposes when work is in progress. In general, persons who may be at increased risk of acquiring infection or for whom infection might be unusually hazardous are not allowed in the animal room.

5. The laboratory or animal facility director establishes policies and procedures whereby only persons who have been advised of the potential hazard and meet any specific requirements (e.g., for immunization) may enter the animal room.

6. When the infectious agent(s) in use in the animal room requires special entry provisions (e.g., vaccination), a hazard warning sign, incorporating the universal biohazard symbol, is posted on the access door to the animal room. The hazard warning sign identifies the infectious agent, lists the name and telephone number of the animal facility supervisor or other responsible person(s), and indicates the special requirement(s) for entering the animal room.

7. Special care is taken to avoid skin contamination with infectious materials; gloves should be worn when handling infected animals and when skin contact with infectious materials is unavoidable.

8. All wastes from the animal room are appropriately decontaminated, preferably by autoclaving, before disposal. Infected animal carcasses are incinerated after being transported from the animal room in leakproof, covered containers.

9. Hypodermic needles and syringes are used only for the parenteral injection or aspiration of fluids from laboratory animals and dia-

phragm bottles. Only needle-locking syringes or disposable needle-syringe units (i.e., the needle is integral to the syringe) are used for the injection or aspiration of infectious fluids. Needles should not be bent, sheared, replaced in the sheath or guard, or removed from the syringe following use. The needle and syringe should be promptly placed in a puncture-resistant container and decontaminated, preferably by autoclaving, before discard or reuse.

10. If floor drains are provided, the drain traps are always filled with water or a suitable disinfectant.

11. When appropriate, considering the agents handled, base-line serum samples from animal care and other at-risk personnel are collected and stored. Additional serum samples may be collected periodically, depending on the agents handled or the function of the facility.

C. *Containment equipment*

Biological safety cabinets, other physical containment devices, and/or personal protective devices (e.g., respirators, face shields) are used whenever procedures with a high potential for creating aerosols are conducted (82). These include necropsy of infected animals, harvesting of infected tissues or fluids from animals or eggs, intranasal inoculation of animals, and manipulations of high concentrations or large volumes of infectious materials.

D. *Animal facilities*

1. The animal facility is designed and constructed to facilitate cleaning and housekeeping.

2. A handwashing sink is available in the room where infected animals are housed.

3. If the animal facility has windows that open they are fitted with fly screens.

4. It is recommended, but not required, that the direction of airflow in the animal facility is inward and that exhaust air is discharged to the outside without being recirculated to other rooms.

5. An autoclave which can be used for decontaminating infectious laboratory waste is available in the building with the animal facility.

Animal Biosafety Level 3

A. *Standard practices*

1. Doors to animals rooms open inward, are self-closing, and are kept closed when work with infected animals is in progress.

2. Work surfaces are decontaminated after use or spills of viable materials.

3. Eating, drinking, smoking, and storing of food for human use are not permitted in the animal room.

4. Personnel wash their hands after handling cultures and animals and before leaving the laboratory.

5. All procedures are carefully performed to minimize the creation of aerosols.

6. An insect and rodent control program is in effect.

B. *Special practices*

1. Cages are autoclaved before bedding is removed and before they are cleaned and washed.

2. Surgical-type masks or other respiratory protection devices (e.g., respirators) are worn by personnel entering rooms housing animals infected with agents assigned to Biosafety Level 3.

3. Wrap-around or solid-front gowns or uniforms are worn by personnel entering the animal room. Front-button laboratory coats are unsuitable. Protective gowns must remain in the animal room and must be decontaminated before being laundered.

4. The laboratory director or other responsible person restricts access to the animal room to personnel who have been advised of the potential hazard and who need to enter the room for program or service purposes when infected animals are present. In general, persons who may be at increased risk of acquiring infection or for whom infection might be unusually hazardous are not allowed in the animal room.

5. The laboratory director or other responsible person establishes policies and procedures whereby only persons who have been advised of the potential hazard and meet any specific re-

quirements (e.g., for immunization) may enter the animal room.

6. Hazard warning signs, incorporating the universal biohazard warning symbol, are posted on access doors to animal rooms containing animals infected with agents assigned to Biosafety Level 3. The hazard warning sign should identify the agent(s) in use, list the name and telephone number of the animal room supervisor or other responsible person(s), and indicate any special conditions of entry into the animal room (e.g., the need for immunizations or respirators).

7. Personnel wear gloves when handling infected animals. Gloves are removed aseptically and autoclaved with other animal room wastes before being disposed of or reused.

8. All wastes from the animal room are autoclaved before disposal. All animal carcasses are incinerated. Dead animals are transported from the animal room to the incinerator in leakproof covered containers.

9. Hypodermic needles and syringes are used only for gavage or for parenteral injection or aspiration of fluids from laboratory animals and diaphragm bottles. Only needle-locking syringes or disposable needle-syringe units (e.g., the needle is integral to the syringe) are used. Needles should not be bent, sheared, replaced in the sheath or guard, or removed from the syringe following use. The needle and syringe should be promptly placed in a puncture-resistant container and decontaminated, preferably by autoclaving, before discard or reuse. Whenever possible, cannulas should be used instead of sharp needles (e.g., for gavage).

10. If floor drains are provided, the drain traps are always filled with water or a suitable disinfectant.

11. If vacuum lines are provided, they are protected with HEPA filters and liquid disinfectant traps.

12. Boots, shoe covers, or other protective footwear and disinfectant footbaths are available and used when indicated.

C. *Containment equipment*

1. Personal protective clothing and equipment and/or other physical containment devices are used for all procedures and manipulations of infectious materials or infected animals.

2. The risk of infectious aerosols from infected animals or their bedding can be reduced if animals are housed in partial containment caging systems, such as open cages placed in ventilated enclosures (e.g., laminar flow cabinets), solid-wall and -bottom cages covered by filter bonnets, or other equivalent primary containment systems.

D. *Animal facilities*

1. The animal facility is designed and constructed to facilitate cleaning and housekeeping and is separated from areas which are open to unrestricted personnel traffic within the building. Passage through two sets of doors is the basic requirement for entry into the animal room from access corridors or other contiguous areas. Physical separation of the animal room from access corridors or other activities may also be provided by a double-doored clothes change room (showers may be included), airlock, or other access facility which requires passage through two sets of doors before entering the animal room.

2. The interior surfaces of walls, floors, and ceilings are water resistant so that they may be easily cleaned. Penetrations in these surfaces are sealed or capable of being sealed to facilitate fumigation or space decontamination.

3. A foot-, elbow-, or automatically operated handwashing sink is provided near each animal room exit door.

4. Windows in the animal room are closed and sealed.

5. Animal room doors are self-closing and are kept closed when infected animals are present.

6. An autoclave for decontaminating wastes is available, preferably within the animal room. Materials to be autoclaved outside the animal room are transported in a covered leakproof container.

7. An exhaust air ventilation system is provided. This system creates directional airflow that draws air into the animal room through the entry area. The building exhaust can be used for

this purpose if the exhaust air is not recirculated to any other area of the building, is discharged to the outside, and is dispersed away from occupied areas and air intakes. Personnel must verify that the direction of the airflow (into the animal room) is proper. The exhaust air from the animal room that does not pass through biological safety cabinets or other primary containment equipment can be discharged to the outside without being filtered or otherwise treated.

8. The HEPA-filtered exhaust air from Class I or Class II biological safety cabinets or other primary containment devices is discharged directly to the outside or through the building exhaust system. Exhaust air from these primary containment devices may be recirculated within the animal room if the cabinet is tested and certified at least every 12 months. If the HEPA-filtered exhaust air from Class I or Class II biological safety cabinets is discharged to the outside through the building exhaust system, it is connected to this system in a manner (e.g., thimble unit connection [80]) that avoids any interference with the air balance of the cabinets or building exhaust system.

Animal Biosafety Level 4

A. *Standard practices*

1. Doors to animal rooms open inward and are self-closing.

2. Work surfaces are decontaminated after use or spills of viable materials.

3. Eating, drinking, smoking, and storing of food for human use are not permitted in the animal room.

4. All procedures are carefully performed to minimize the creation of aerosols.

5. An insect and rodent control program is in effect.

6. Cages are autoclaved before bedding is removed and before they are cleaned and washed.

B. *Special practices*

1. Only persons whose entry into the facility or individual animal rooms is required for program or support purposes are authorized to enter. Persons who may be at increased risk of acquiring infection or for whom infection might be unusually hazardous are not allowed in the animal facility. Persons at increased risk may include children, pregnant women, and persons who are immunodeficient or immunosuppressed. The supervisor has the final responsibility for assessing each circumstance and determining who may enter or work in the laboratory. Access to the facility is limited by secure, locked doors; accessibility is controlled by the animal facility supervisor, biohazards control officer, or other person responsible for the physical security of the facility. Before entering, persons are advised of the potential biohazards and instructed as to appropriate safeguards. Personnel comply with the instructions and all other applicable entry and exit procedures. Practical and effective protocols for emergency situations are established.

2. Personnel enter and leave the facility only through the clothing change and shower rooms. Personnel shower each time they leave the facility. Head covers are provided to personnel who do not wash their hair during the exit shower. Except in an emergency, personnel do not enter or leave the facility through the airlocks.

3. Street clothing is removed in the outer clothing change room and kept there. Complete laboratory clothing, including undergarments, pants and shirts or jumpsuits, shoes, and gloves, is provided and used by all personnel entering the facility. When exiting, personnel remove laboratory clothing and store it in a locker or hamper in the inner change room before entering the shower area.

4. When infectious materials or infected animals are present in the animal rooms, a hazard warning sign, incorporating the universal biohazard symbol, is posted on all access doors. The sign identifies the agent, lists the name and telephone number of the animal facility supervisor or other responsible person(s), and indicates any special conditions of entry into the area (e.g., the need for immunizations and respirators).

5. Supplies and materials to be taken into the facility enter by way of the double-door autoclave, fumigation chamber, or airlock, which is appropriately decontaminated between each use. After securing the outer doors, personnel inside

the facility retrieve the materials by opening the interior doors of the autoclave, fumigation chamber, or airlock. This inner door is secured after materials are brought into the facility.

6. Materials (e.g., plants, animals, clothing) not related to the experiment are not permitted in the facility.

7. Hypodermic needles and syringes are used only for gavage or for parenteral injection and aspiration of fluids from laboratory animals and diaphragm bottles. Only needle-locking syringes or disposable syringe-needle units (i.e., needle is integral part of unit) are used. Needles should not be bent, sheared, replaced in the guard or sheath, or removed from the syringe following use. The needle and syringe should be promptly placed in a puncture-resistant container and decontaminated, preferably by autoclaving, before discard or reuse. Whenever possible, cannulas should be used instead of sharp needles (e.g., for gavage).

8. A system is developed and is operational for the reporting of animal facility accidents and exposures, employee absenteeism, and for the medical surveillance of potential laboratory-associated illnesses. An essential adjunct to such a reporting-surveillance system is the availability of a facility for the quarantine, isolation, and medical care of persons with potential or known laboratory-associated illnesses.

9. Base-line serum samples are collected and stored for all laboratory and other at-risk personnel. Additional serum specimens may be collected periodically, depending on the agents handled or the function of the laboratory.

C. *Containment equipment*

Laboratory animals infected with agents assigned to Biosafety Level 4 are housed in a Class III biological safety cabinet or in partial-containment caging systems (such as open cages placed in ventilated enclosures, solid-wall and -bottom cages covered with filter bonnets, or other equivalent primary containment systems) in specially designed areas in which all personnel are required to wear one-piece positive-pressure suits ventilated with a life support system. Animal work with viral agents that require Biosafety Level 4 secondary containment and for which highly effective vaccines are available and used may be conducted with partial-containment cages and without the one-piece positive-pressure personnel suit if the facility has been decontaminated, if no concurrent experiments are being done in the facility which require Biosafety Level 4 primary and secondary containment, and if all other standard and special practices are followed.

D. *Animal facility*

1. The animal rooms are located in a separate building or in a clearly demarcated and isolated zone within a building. Outer and inner change rooms separated by a shower are provided for personnel entering and leaving the facility. A double-doored autoclave, fumigation chamber, or ventilated airlock is provided for passage of materials, supplies, or equipment which are not brought into the facility through the change room.

2. Walls, floors, and ceilings of the facility are constructed to form a sealed internal shell which facilitates fumigation and is animal and insect proof. The internal surfaces of this shell are resistant to liquids and chemicals, thus facilitating cleaning and decontamination of the area. All penetrations in these structures and surfaces are sealed.

3. Internal facility appurtenances, such as light fixtures, air ducts, and utility pipes, are arranged to minimize the horizontal surface area on which dust can settle.

4. A foot-, elbow-, or automatically operated handwashing sink is provided near the door of each animal room within the facility.

5. If there is a central vacuum system, it does not serve areas outside of the facility. The vacuum system has in-line HEPA filters placed as near as practicable to each use point or service cock. Filters are installed to permit in-place decontamination and replacement. Other liquid and gas services for the facility are protected by devices that prevent backflow.

6. External animal facility doors are self-closing and self-locking.

7. Any windows must be resistant to breakage and sealed.

8. A double-doored autoclave is provided for decontaminating materials that leave the facility. The autoclave door which opens to the

area external to the facility is automatically controlled so that it can only be opened after the autoclave "sterilization" cycle is completed.

9. A pass-through dunk tank, fumigation chamber, or an equivalent decontamination method is provided so that materials and equipment that cannot be decontaminated in the autoclave can be safely removed from the facility.

10. Liquid effluents from laboratory sinks, cabinets, floors, and autoclave chambers are decontaminated by heat treatment before being discharged. Liquid wastes from shower rooms and toilets may be decontaminated with chemical disinfectants or by heat in the liquid waste decontamination system. The procedure used for heat decontamination of liquid wastes must be evaluated mechanically and biologically by using a recording thermometer and an indicator microorganism with a defined heat susceptibility pattern. If liquid wastes from the shower rooms are decontaminated with chemical disinfectants, the chemicals used must have documented efficacy against the target or indicator microorganisms.

11. An individual supply and exhaust air ventilation system is provided. The system maintains pressure differentials, and directional airflow is required to assure inflow from areas outside the facility toward areas of highest potential risk within the facility. Manometers are provided to sense pressure differentials between adjacent areas and are maintained at different pressure levels. The manometers sound an alarm when a system malfunctions. The supply and exhaust airflow is interlocked to assure inward (or zero) airflow at all times.

12. Air can be recirculated within an animal room if it is filtered through a HEPA filter.

13. The exhaust air from the facility is filtered by HEPA filters and discharged to the outside so that it is dispersed away from occupied buildings and air intakes. Within the facility, the filters are located as near to the laboratories as practicable in order to reduce the length of potentially contaminated air ducts. The filter chambers are designed to allow in situ decontamination before filters are removed and to facilitate certification testing after they are replaced. Coarse filters are provided for treatment of air supplied to the facility in order to increase the lifetime of the HEPA filters.

14. The treated exhaust air from Class I or Class II biological safety cabinets can be discharged into the animal room environment or to the outside through the facility air exhaust system. If exhaust air from Class I or II biological safety cabinets is discharged into the animal room, the cabinets are tested and certified at 6-month intervals. *The treated exhaust air from Class III biological safety cabinets is discharged without recirculation via the facility exhaust air system.* If the treated exhaust air from any of these cabinets is discharged to the outside through the facility exhaust air system, it is connected to this system in a manner that avoids any interference with the air balance of the cabinets or the facility exhaust air system.

15. A specially designed suit area may be provided in the facility. Personnel who enter this area wear a one-piece positive-pressure suit that is ventilated by a life support system. The life support system is provided with alarms and emergency backup breathing air tanks. Entry to this area is through an airlock fitted with airtight doors. A chemical shower is provided to decontaminate the surface of the suit before the worker leaves the area. The exhaust air from the area in which the suit is used is filtered by two sets of HEPA filters installed in series. A duplicate filtration unit and exhaust fan are provided. An automatically starting emergency power source is provided. The air pressure within the suit area is lower than that of any adjacent area. Emergency lighting and communication systems are provided. All penetrations into the inner shell of the suit area are sealed. A double-doored autoclave is provided for decontaminating waste materials to be removed from the suit area.

SECTION V. RECOMMENDED BIOSAFETY LEVELS FOR INFECTIOUS AGENTS AND INFECTED ANIMALS

Selection of an appropriate biosafety level for work with a particular agent or animal study depends upon a number of factors. Some of the most important are: the virulence, pathogenicity, biological stability, route of spread, and communicability of the agent; the nature or function of the laboratory; the procedures and manipulations involving the agent; the quantity and concentration of the agent; the endemicity of the agent; and the availability of effective vaccines or therapeutic measures.

Agent summary statements in this section provide guidance for the selection of appropriate biosafety levels. Specific information on laboratory hazards associated with a particular agent and recommendations regarding practical safeguards that can significantly reduce the risk of laboratory-associated diseases are included. Agent summary statements are presented for agents which meet one or more of the following criteria: the agent is a proven hazard to laboratory personnel working with infectious materials (e.g., hepatitis B virus, tubercle bacilli); the potential for laboratory-associated infection is high even in the absence of previously documented laboratory-associated infections (e.g., exotic arboviruses); or the consequences of infection are grave (e.g., Creutzfeldt-Jakob disease, botulism).

Recommendations for the use of vaccines and toxoids are included in agent summary statements when such products are available—either as licensed or Investigational New Drug (IND) products. When applicable, recommendations for the use of these products are based on current recommendations of the Public Health Service Advisory Committee on Immunization Practice and are specifically targeted to at-risk laboratory personnel and others who must work in or enter laboratory areas. These specific recommendations should in no way preclude the routine use of such products as diphtheria-tetanus toxoids, poliovirus vaccine, influenza vaccine, and others because of the potential risk of community exposures irrespective of any laboratory risks. Appropriate precautions should be taken in the administration of live attenuated virus vaccines in individuals with altered immunocompetence.

Risk assessments and Biosafety Levels recommended in the agent summary statements presuppose a population of immunocompetent individuals. Those with altered immunocompetence may be at increased risk when exposed to infectious agents. Immunodeficiency may be hereditary, congenital, or induced by a number of neoplastic diseases, by therapy, or by radiation. The risk of becoming infected or the consequences of infection may also be influenced by such factors as age, sex, race, pregnancy, surgery (e.g., splenectomy, gastrectomy), predisposing diseases (e.g., diabetes, lupus erythematosus), or altered physiological function. These and other variables must be considered in individualizing the generic risk assessments of the agent summary statements for specific activities.

The basic biosafety level assigned to an agent is based on the activities typically associated with the growth and manipulation of quantities and concentrations of infectious agents required to accomplish identification of typing. If activities with clinical materials pose a lower risk to personnel than those activities associated with manipulation of cultures, a lower biosafety level is recommended. On the other hand, if the activities involve large volumes or highly concentrated preparations ("production quantities") or manipulations which are likely to produce aerosols or which are otherwise intrinsically hazardous, additional personnel precautions and increased levels of primary and secondary containment may be indicated. "Production quantities" refers to large volumes or concentrations of infectious agents considerably in excess of those typically used in identification and typing activities. Propagation and concentration of infectious agents, as occurs in large-scale fermentations, antigen and vaccine production, and a variety of other commercial and research activities, clearly deal with significant masses of infectious agents that are reasonably considered "production quantities." However, in terms of potentially increased risk as a function of the mass of infectious agents, it is not possible to define "production quantities" in finite volumes or concentrations for any given agent. Therefore, the laboratory director must make a risk assessment of the activities conducted and select practices, containment equipment, and facilities appropriate to the risk, irrespective of the volume or concentration of agent involved.

Occasions will arise when the laboratory director should select a biosafety level higher than that recommended. For example, a higher biosafety level may be indicated by the unique nature of the proposed activity (e.g., the need for special containment for experimentally generated aerosols for inhalation studies) or by the proximity of the laboratory to areas of special concern (e.g., a diagnostic laboratory located near patient care areas). Similarly, a recommended biosafety level may be adapted to compensate for the absence of certain recommended safeguards. For example, in those situations where Biosafety Level 3 is recommended, acceptable safety may be achieved for routine or repetitive operations (e.g., diagnostic procedures involving the propagation of an agent for identification, typing, and susceptibility testing) in laboratories where facility features satisfy Biosafety Level 2 recommendations, provided the recommended "Standard Microbiological Practices," "Special Practices," and "Containment Equipment" for Biosafety Level 3 are rigorously followed. The decision to adapt Biosafety Level 3 recommendations in this manner should be made only by the laboratory director. This adaptation, however, is not suggested for agent production operations or activities where procedures are frequently changing. The laboratory director should also give special consideration to selecting appropriate safeguards for materials that may contain a suspected agent. For example, sera of human origin may contain hepatitis B virus and should be handled under conditions which reasonably preclude cutaneous, mucous membrane, or parenteral exposure of personnel, and sputa submitted to the laboratory for assay for tubercle bacilli should be handled under conditions which reasonably preclude the generation of aerosols or which contain any aerosols that may be generated during the manipulation of clinical materials or cultures.

The infectious agents which meet the previously stated criteria are listed by category of agent on the following pages. To use these summaries, first locate the agent in the listing under the appropriate category of agent. Second, utilize the practices, safety equipment, and type of facilities recommended for working with clinical materials, cultures of infectious agents, or infected animals recommended in the agent summary statement and described in Section V.

The laboratory director is also responsible for appropriate risk assessment of agents not included in the Agent Summary Statements and for utilization of appropriate practices, containment equipment, and facilities for the agent used.

Risk assessment. The risk assessment of laboratory activities involving the use of infectious microorganisms is ultimately a subjective process. Those risks associated with the agent, as well as with the activity to be conducted, must be considered in the assessment. The characteristics of infectious agents and the primary laboratory hazards of working with the agent are described generically for agents in Biosafety Levels 1-4 and specifically for individual agents or groups of agents on pages 87-88 and in Section V, respectively, of this Appendix.

Hepatitis B virus (HBv) is an appropriate model for illustrating the risk assessment process. HBv is among the most ubiquitous of human pathogens and most prevalent of laboratory-associated infections. The agent has been demonstrated in a variety of body secretions and excretions. Blood, saliva, and semen have been shown to be infectious. Natural transmission is associated with parenteral inoculation or with contamination of the broken skin or of mucous membranes with infectious body fluids. There is no evidence of airborne or interpersonal spread through casual contact. Prophylactic measures include the use of a licensed vaccine in high-risk groups and the use of hepatitis B immune globulin following overt exposure.

The primary risk of HBv infection in laboratory personnel is associated with accidental parenteral inoculation, exposure of the broken skin or mucous membranes of the eyes, nose, or mouth, or ingestion of infectious body fluids. These risks are typical of those described for Biosafety Level 2 agents and are addressed by using the recommended standard and special microbiological practices to minimize or eliminate these overt exposures.

Hepatitis non-A non-B and AIDS—acquired immune deficiency syndrome—pose similar infection risks to laboratory personnel. The prudent practices recommended for HBv are applicable to these two disease entities, as well as to the routine laboratory manipulation of clinical materials of domestic origin.

The described risk assessment process is also applicable to laboratory operations other than those involving the use of primary agents of human disease. Microbiological studies of animal host-specific pathogens, soil, water, food, feeds, and other natural or manufactured materials, by comparison, pose substantially lower risks of laboratory infection. Microbiologists and other scientists working with such materials may nevertheless find the practices, containment equipment, and facilities recommendations described in this publication of value in developing operational standards to meet their own assessed needs.

AGENT SUMMARY STATEMENTS

Parasitic Agents

Agent: **Nematode parasites of humans**

Laboratory-associated infections with *Strongyloides* spp. and hookworms have been reported (90). Allergic reactions to various antigenic components of nematodes (e.g., aerosolized *Ascaris* antigens) may represent an individual risk to sensitized persons. Laboratory animal-associated infections (including arthropods) have not been reported, but infective larvae in the feces of nonhuman primates and of dogs infected with *Strongyloides* spp. are a potential infection hazard for laboratory and animal care personnel.

Laboratory hazards. Eggs and larvae in freshly passed feces of infected hosts are usually not infective; development of the infective stages may take periods of 1 day to several weeks. Ingestion of the infective eggs or skin penetration of infective larvae are the primary hazards to laboratory and animal care personnel. Arthropods infected with filarial parasites pose a potential hazard to laboratory personnel. In laboratory personnel with frequent exposure to aerosolized antigens of *Ascaris* spp., development of hypersensitivity is common.

Recommended precautions. Biosafety Level 2 practices, containment equipment, and facilities are recommended for activities with infective stages of the parasites listed. Exposure to aerosolized sensitizing antigens of *Ascaris* spp. should be avoided. Primary containment (e.g., biological safety cabinet) may be required for work with these materials by hypersensitive individuals.

Agent: **Protozoal parasites of humans**

Laboratory-associated infections with *Toxoplasma* spp., *Plasmodium* spp. (including *P. cynomologi*), *Trypanosoma* spp., and *Leishmania* spp. have been reported (21, 49, 90, 100). In addition, infections with *Entamoeba histolytica*, *Giardia* spp., and *Coccidia* spp. can result from ingestion of cysts in feces.

Accidental laboratory infections as well as human volunteer studies have proven the transmissibility of *P. cynomologi* from nonhuman primates to humans via infected mosquitoes (40). Although laboratory animal-associated infections have not been reported, contact with lesion material from rodents with cutaneous leishmaniasis and with feces or blood of experimentally or naturally infected animals may be a direct source of infection for laboratory personnel.

Laboratory hazards. Infective stages may be present in blood, feces, lesion exudates, and infected arthropods. Depending on the parasite, accidental parenteral inoculation, transmission by arthropod vectors, skin penetration, and ingestion are the primary laboratory hazards. Aerosol or droplet exposure of the mucous membranes of the eyes, nose, or mouth with trophozoites are potential hazards when working with cultures of *Naegleria fowleri*, *Leishmania* spp., *Trypanosoma cruzi*, or tissue homogenates or blood containing hemoflagellates. Because of the grave consequence of toxoplasmosis in the developing fetus, women of childbearing age should be discouraged from working with viable *Toxoplasma* spp.

Recommended precautions. Biosafety Level 2 practices, containment equipment, and facilities are recommended for activities with *infective stages* of the parasites listed. Infected arthropods should be maintained in facilities which reasonably preclude the exposure of personnel or their escape to the outside. Primary containment (e.g., biological safety cabinet) or personal protection (e.g., face shield) may be indicated when working with cultures of *T. cruzi*, *Leishmania*, *N. fowleri*, or tissue homogenates or blood containing hemoflagellates. Gloves are recommended for activities where there is the likelihood of direct skin contact with infective stages of the parasites listed.

Agent: **Trematode parasites of humans**

Laboratory-associated infections with *Schistosoma* spp. and *Fasciola* spp. have been reported—none associated directly with laboratory animals (90).

Laboratory hazards. Infective stages of *Schistosoma* spp. (cercariae) and *Fasciola* spp. (metacercariae) may be found, respectively, in the water or encysted on aquatic plants in laboratory aquaria used to maintain snail intermediate hosts. Skin penetration by schistosome cercariae and ingestion of fluke metacercariae are the primary laboratory hazards. Dissection or crushing of schistosome-infected snails may also result in exposure of skin or mucous membranes to cercariae-containing droplets. Additionally, metacercariae may be inadvertently transferred from hand to mouth by fingers or gloves following contact with contaminated aquatic vegetation or surfaces of aquaria. Most laboratory exposure to *Schistosoma* spp. would predictably result in low worm burdens with minimal disease potential. Safe and effective drugs are available for the treatment of schistosomiasis.

Recommended precautions. Biosafety Level 2 practices, containment equipment, and facilities are recommended for activities with *infective stages* of the parasites listed. Gloves should be worn when there may be direct contact with water containing cercariae or vegetation containing metacercariae from naturally or experimentally infected snail intermediate hosts. Snails and cercariae in the water of laboratory aquaria should be killed by chemicals (e.g., hypochlorites, iodine) or heat before discharge to sewers.

Agent: **Cestode parasites of humans**

Although laboratory-associated infections with *Echinococcus granulosus* or *Taenia solium* have not been reported, the consequences of such infections following the ingestion of infective eggs of *T. solium* or *E. granulosus* are potentially grave.

Laboratory hazards. Infective eggs may be present in the feces of dogs or other canids (the definitive hosts of *E. granulosus*) or in the feces of humans (the definitive host of *T. solium*). Ingestion of infective eggs from these sources is the primary laboratory hazard. Cysts and cyst fluids of *E. granulosus* are not infectious for humans.

Recommended precautions. Biosafety Level 2 practices, containment equipment, and facilities are recommended for work with infective stages of these parasites. Special attention should be given to personal hygiene practices (e.g., handwashing) and avoidance of ingestion of infective eggs. Gloves are recommended when there may be direct contact with feces or surfaces contaminated with fresh feces of dogs infected with *E. granulosus* or humans infected with *T. solium* adults.

Fungal Agents

Agent: ***Blastomyces dermatitidis***

Laboratory-associated local infections following accidental parenteral inoculation with infected tissues or cultures containing yeast forms of *B. dermatitidis* (39, 54, 66, 103, 127) have been reported. A single pulmonary infection (asymptomatic) occurred following the presumed inhalation of conidia. Subsequently this individual developed an osteolytic lesion from which *B. dermatitidis* was cultured (30). Presumably, pulmonary infections are associated only with sporulating mold forms (conidia).

Laboratory hazards. Yeast forms may be present in the tissues of infected animals and in clinical specimens. Parenteral (subcutaneous) inoculation of these materials may cause local granulomas. Mold form cultures of *B. dermatitidis* containing infectious conidia may pose a hazard of aerosol exposure.

Recommended precautions. Biosafety Level 2 and Animal Biosafety Level 2 practices, containment equipment, and facilities are recommended for activities with clinical materials, animal tissues, and infected animals.

Biosafety Level 3 practices, containment equipment, and facilities are recommended for processing mold cultures, soil, and other environmental materials known or likely to contain infectious conidia.

Agent: ***Coccidioides immitis***

Laboratory-associated coccidioidomycosis is a documented hazard (12, 28, 31, 32, 33, 64, 68, 79, 105, 106, 107). Wilson et al. reported that 28 of 31 (90%) laboratory-associated infections in his institution resulted in clinical disease, whereas more than half of infections acquired in nature were asymptomatic (128).

Laboratory hazards. Because of its size (2 to 5 μm), the arthrospore is conducive to ready dispersal in air and retention in the deep pulmonary spaces. The much larger size of the spherule (30 to 60 μm) considerably reduces the effectiveness of this form of the fungus as an airborne pathogen.

Spherules of the fungus may be present in clinical specimens and animal tissues, and infectious arthrospores may be present in mold cultures and soil samples. Inhalation of arthrospores from soil samples or mold cultures or following transformation from the spherule form in clinical materials is the primary laboratory hazard. Accidental percutaneous inoculation of the spherule form may result in local granuloma formation (118). Disseminated disease may occur at a greater frequency in pregnant women, blacks, and Filipinos than in whites.

Recommended precautions. Biosafety Level 2 practices, containment equipment, and facilities are

recommended for handling and processing clinical specimens and animal tissues. Animal Biosafety Level 2 practices and facilities are recommended for experimental animal studies when the route of challenge is parenteral.

Biosafety Level 3 practices and facilities are recommended for all activities with sporulating mold form cultures of *C. immitis* and for processing soil or other environmental materials known or likely to contain infectious arthrospores.

Agent: ***Cryptococcus neoformans***

A single account of a laboratory exposure to *C. neoformans* as a result of a laceration by a scalpel blade heavily contaminated with encapsulated cells is reported (50). This vigorous exposure, which did not result in local or systemic evidence of infection, suggests that the level of pathogenicity for normal immunocompetent adults is low. Respiratory infections as a consequence of laboratory exposure have not been recorded.

Laboratory hazards. Accidental parenteral inoculation of cultures or other infectious materials represents a potential hazard to laboratory personnel—particularly to those that may be immunocompromised. Bites by experimentally infected mice and manipulations of infectious environmental materials (e.g., pigeon droppings) may also represent a potential hazard to laboratory personnel.

Recommended precautions. Biosafety Level 2 and Animal Biosafety Level 2 practices, containment equipment, and facilities are recommended, respectively, for activities with known or potentially infectious clinical, environmental, or culture materials and with experimentally infected animals.

The processing of soil or other environmental materials known or likely to contain infectious yeast cells should be conducted in a Class I or Class II biological safety cabinet. This precaution is also indicated for cultures of the perfect or sexual state of the agent.

Agent: ***Histoplasma capsulatum***

Laboratory-associated histoplasmosis is a documented hazard in facilities conducting diagnostic or investigative work (90, 91). Pulmonary infections have resulted from handling mold form cultures (78). Local infection has resulted from skin puncture during autopsy of an infected human (119) and from accidental needle inoculation of a viable culture (116). Collecting and processing soil samples from endemic areas have caused pulmonary infections in laboratory workers. Spores are resistant to drying and may remain viable for long periods of time. The small size of the infective conidia (microconidia are less than 5 µm) is conducive to airborne dispersal and intrapulmonary retention. Furcolow reported that 10 spores were almost as effective as a lethal inoculum in mice as 10,000 to 100,000 spores (45).

Laboratory hazards. The infective stage of this dimorphic fungus (conidia) is present in sporulating mold form cultures and in soil from endemic areas. The yeast form in tissues or fluids from infected animals may produce local infection following parenteral inoculation.

Recommended precautions. Biosafety Level 2 and Animal Biosafety Level 2 practices, containment equipment, and facilities are recommended for handling and processing clinical specimens and animal tissues and for experimental animal studies when the route of challenge is parenteral.

Biosafety Level 3 practices and facilities are recommended for processing mold cultures, soil, or other environmental materials known or likely to contain infectious conidia.

Agent: ***Sporothrix schenckii***

S. schenckii has caused a substantial number of local skin or eye infections in laboratory personnel. Most cases have been associated with accidents and have involved splashing culture material into the eye (41, 125), scratching (13) or injecting (117) infected material into the skin, or being bitten by an experimentally infected animal (60, 61). Skin infections have resulted also from handling cultures (74, 81) or necropsy of animals (44) without a known break in technique. No pulmonary infections have been reported to result from laboratory exposure, although naturally occurring lung disease, albeit rare, is thought to result from inhalation.

Recommended precautions. Biosafety Level 2 and Animal Biosafety Level 2 practices, containment equipment, and facilities are recommended for all laboratory and experimental animal activities with *S. schenckii*.

Agents: **Pathogenic members of the genera *Epidermophyton, Microsporum,* and *Trichophyton***

Although skin, hair, and nail infections by these dermatophytic molds are among the most prevalent of human infections, the processing of clinical material has not been associated with laboratory infections. Infections have been acquired through contacts with naturally or experimentally infected laboratory animals (mice, rabbits, guinea pigs, etc.) and, rarely, with handling cultures (71, 90, 51).

Laboratory hazards. Agents are present in the skin, hair, and nails of human and animal hosts. Contact with infected laboratory animals with inapparent or apparent infections is the primary hazard to laboratory personnel. Cultures and clinical materials are not an important source of human infection.

Recommended precautions. Biosafety Level 2 and Animal Biosafety Level 2 practices, containment equipment, and facilities are recommended for all laboratory and experimental animal activities with dermatophytes.

Bacterial Agents

Agent: *Bacillus anthracis*

Forty (40) cases of laboratory-associated anthrax, occurring primarily at facilities conducting anthrax research, have been reported (38, 90). No laboratory-associated cases of anthrax have been reported in the United States for more than 20 years.

Naturally and experimentally infected animals pose a potential risk to laboratory and animal care personnel.

Laboratory hazards. The agent may be present in blood, skin lesion exudates, and, rarely, in urine and feces. Direct and indirect contact of the intact and broken skin with cultures and contaminated laboratory surfaces, accidental parenteral inoculation, and, rarely, exposure to infectious aerosols are the primary hazards to laboratory personnel.

Recommended precautions. Biosafety Level 2 practices, containment equipment, and facilities are recommended for activities using clinical materials and diagnostic quantities of infectious cultures. Animal Biosafety Level 2 practices and facilities are recommended for studies utilizing experimentally infected laboratory rodents. A licensed vaccine is available through the Centers for Disease Control; however, vaccination of laboratory personnel is not recommended unless frequent work with clinical specimens or diagnostic cultures is anticipated (e.g., animal disease diagnostic laboratory). Biosafety Level 3 practices and facilities are recommended for work involving production volumes or concentrations of cultures and for activities which have a high potential for aerosol production. In these facilities vaccination is recommended for all persons working with the agent, all persons working in the same laboratory room where the cultures are handled, and persons working with infected animals.

Agent: *Brucella (B. abortus, B. canis, B. melitensis, B. suis)*

B. abortus, B. canis, B. melitensis, and *B. suis* have all caused illness in laboratory personnel (77, 90, 110). Brucellosis is the most commonly reported laboratory-associated bacterial infection (90). Hypersensitivity to *Brucella* antigens is also a hazard to laboratory personnel.

Occasional cases have been attributed to exposure to experimentally and naturally infected animals or their tissues.

Laboratory hazards. The agent may be present in blood, cerebrospinal fluid, semen, and occasionally urine. Most laboratory-associated cases have occurred in research facilities and have involved exposure to *Brucella* organisms being grown in large quantities. Direct skin contact with cultures or with infectious clinical specimens from animals (e.g., blood, uterine discharges) are also commonly implicated. Aerosols generated during laboratory procedures have caused large outbreaks (59). Mouth pipetting, accidental parenteral inoculations, and sprays into eyes, nose, and mouth have also resulted in infection.

Recommended precautions. Biosafety Level 2 practices are recommended for activities with clinical materials of human or animal origin containing or potentially containing pathogenic *Brucella* spp. Biosafety Level 3 and Animal Biosafety Level 3 practices, containment equipment, and facilities are recommended, respectively, for all manipulations of cultures of the pathogenic *Brucella* spp. listed in this summary and for experimental animal studies. Vaccines are not available for use in humans.

Agent: *Chlamydia psittaci, C. trachomatis*

Infections with psittacosis, lymphogranuloma venereum (LGV), and trachoma are documented hazards and the fifth most commonly reported laboratory-associated bacterial infection. The majority of cases were of psittacosis, occurred before 1955, and had the highest case-fatality rate of all groups of infectious agents (90). Contact with and exposure to infectious aerosols in the handling, care, or necropsy of naturally or experimentally infected birds are the major sources of laboratory-associated psittacosis. Infected mice and eggs are less important sources of *C. psittaci*. Laboratory animals are not a reported source of human infection with *C. trachomatis*.

Laboratory hazards. *C. psittaci* may be present in the tissues, feces, nasal secretions, and blood of infected birds and in blood, sputum, and tissues of infected humans. *C. trachomatis* may be present in genital, bubo, and conjunctival fluids of infected humans. Exposure to infectious aerosols and droplets created during the handling of infected birds and tissues is the primary hazard to laboratory personnel working with psittacosis. The primary laboratory hazards of *C. trachomatis* are accidental parenteral inoculation and direct and indirect exposure of mucous membranes of the eyes, nose, and mouth to genital, bubo, or conjunctival fluids, cell culture materials, and fluids from infected eggs. Infectious aerosols may also pose a potential source of infection.

Recommended precautions. Biosafety Level 2 practices, containment equipment, and facilities are recommended for activities involving the necropsy of infected birds and the diagnostic examination of tissues of cultures known to be infected or potentially infected with *C. psittaci* or *C. trachomatis*. Wetting the feathers of infected birds with a detergent-disinfectant prior to necropsy can appreciably reduce the risk of aerosols of infected feces and nasal secretions on the feathers and external surfaces of the bird. Animal Biosafety Level 2 practices and facilities and respiratory protection are recommended for personnel working with caged birds naturally or experimentally infected. Gloves are recommended for the necropsy of birds and mice, the opening of inoculated eggs, and when there is the likelihood of direct skin contact with infected tissues, bubo fluids, and other clinical materials. Additional primary containment and personnel precautions, such as those recommended for Biosafety Level 3, may be indicated for activities with high potential for droplet or aerosol production and for activities involving production quantities or concentrations of infectious materials. Vaccines are not available for use in humans.

Agent: ***Clostridium botulinum***

While there are no reported cases of botulism associated with the handling of the agent or toxin in the laboratory or working with naturally or experimentally infected animals, the consequences of such intoxications would be grave.

Laboratory hazards. *C. botulinum* or its toxin may be present in a variety of food products, clinical materials (serum, feces), and environmental samples (soil, surface water). Exposure to the toxin of *C. botulinum* is the primary laboratory hazard. The toxin may be absorbed after ingestion or following contact with the skin, eyes, or mucous membranes, including the respiratory tract. Accidental parenteral inoculation may also represent a significant exposure to toxin. Broth cultures grown under conditions of optimal toxin production may contain 2×10^6 mouse LD_{50} per ml (111).

Recommended precautions. Biosafety Level 2 practices, containment equipment, and facilities are recommended for all activities with materials known to contain or potentially containing the toxin. A pentavalent (ABCDE) botulism toxoid is available through the Centers for Disease Control as an Investigational New Drug (IND). This toxoid is recommended for personnel working with cultures of *C. botulinum* or its toxins. Solutions of sodium hydroxide (0.1 N) readily inactivate the toxin and are recommended for decontaminating work surfaces and spills of cultures or toxin. Additional primary containment and personnel precautions, such as those recommended for Biosafety Level 3, may be indicated for activities with a high potential for aerosol or droplet production, those involving production quantities of toxin, and those involving purified toxins. Animal Biosafety Level 2 practices and facilities are recommended for diagnostic studies and titration of toxin.

Agent: ***Clostridium tetani***

Although the risk of infection to laboratory personnel is negligible, Pike (90) has recorded five incidents related to exposure of personnel during manipulation of the toxin.

Laboratory hazards. Accidental parenteral inoculation and ingestion of the toxin are the primary hazards to laboratory personnel. Since tetanus toxin is poorly absorbed through mucous membranes, aerosols and droplets probably represent minimal hazards.

Recommended precautions. Biosafety Level 2 practices, containment equipment, and facilities are recommended for activities involving the manipulation of cultures or toxin. While the risk of laboratory-associated tetanus is low, the administration of an adult diphtheria-tetanus toxoid at 10-year intervals may further reduce the risk to laboratory and animal care personnel of toxin exposures and wound contamination (24).

Agent: ***Corynebacterium diphtheriae***

Laboratory-associated infections with *C. diphtheriae* are documented. Pike (90) lists 33 cases reported in the world literature.

Laboratory animal-associated infections have not been reported.

Laboratory hazards. The agent may be present in exudates or secretions of the nose, throat (tonsil), pharynx, larynx, and wounds, in blood, and on the skin. Inhalation, accidental parenteral inoculation, and ingestion are the primary laboratory hazards.

Recommended precautions. Biosafety Level 2 practices, containment equipment, and facilities are recommended for all activities utilizing known or potentially infected clinical materials or cultures. Animal Biosafety Level 2 facilities are recommended for studies utilizing infected laboratory animals. While the risk of laboratory-associated diphtheria is low, the administration of an adult diphtheria-tetanus toxoid at 10-year intervals may further reduce the risk to laboratory and animal care personnel of toxin exposures and work with infectious materials (24).

Agent: ***Francisella tularensis***

Tularemia is the third most commonly reported laboratory-associated bacterial infection (90). Almost all cases occurred at facilities involved in tularemia research. Occasional cases have been related to work with naturally or experimentally infected animals or their ectoparasites.

Laboratory hazards. The agent may be present in lesion exudate, respiratory secretions, cerebrospinal fluid, blood, urine, tissues from infected animals, and fluids from infected arthropods. Direct contact of skin or mucous membranes with infectious materials, accidental parenteral inoculation, ingestion, and exposure to aerosols and infectious droplets have resulted in infection. Cultures have been more commonly associated with infection than have clinical materials and infected animals. The human 25-50% infectious dose is on the order of 10 organisms by the respiratory route (121).

Recommended precautions. Biosafety Level 2 practices, containment equipment, and facilities are recommended for activities with clinical materials of human or animal origin containing or potentially containing *F. tularensis*. Biosafety Level 3 and Animal Biosafety Level 3 practices and facilities are recommended, respectively, for all manipulations of cultures and for experimental animal studies. An investigational live attenuated vaccine (10) is available through the Centers for Disease Control and is recommended for persons working with the agent or with infected animals and for persons working in or entering the laboratory or animal room where cultures or infected animals are maintained.

Agent: ***Leptospira interrogans*—all serovars**

Leptospirosis is a well-documented laboratory hazard. Sixty-seven laboratory-associated infections and 10 deaths have been reported (90).

An experimentally infected rabbit was identified as the source of an infection with *L. interrogans* serovar *icterohemorrhagiae* (97). Direct and indirect contact with fluids and tissues of experimentally or naturally infected mammals during handling, care, or necropsy are potential sources of infection. In animals with chronic kidney infections, the agent is shed in the urine in enormous numbers for long periods of time.

Laboratory hazards. The agent may be present in urine, blood, and tissues of infected animals and humans. Ingestion, accidental parenteral inoculation, and direct and indirect contact of skin or mucous membranes with cultures or infected tissues or body fluids—especially urine—are the primary laboratory hazards. The importance of aerosol exposure is not known.

Recommended precautions. Biosafety Level 2 practices, containment equipment, and facilities are recommended for all activities involving the use or manipulation of known or potentially infectious tissues, body fluids, and cultures and for the housing of infected animals. Gloves are recommended for the handling and necropsy of infected animals and when there is the likelihood of direct skin contact with infectious materials. Vaccines are not available for use in humans.

Agent: ***Legionella pneumophila*; other *Legionella*-like agents**

A single documented nonfatal laboratory-associated case of legionellosis due to presumed aerosol or droplet exposure during animal challenge studies with Pontiac fever agent (*L. pneumophila*) is recorded (16). Human-to-human spread has not been documented.

Experimental infections are readily produced in guinea pigs and embryonated chicken eggs (72). Challenged rabbits develop antibodies but not clinical disease. Mice are refractory to parenteral exposure. Unpublished studies by Kaufmann, Feeley, and others at the Centers for Disease Control have shown that animal-to-animal transmission did not occur in a variety of experimentally infected mammalian and avian species.

Laboratory hazards. The agent may be present in pleural fluids, tissue, sputa, and environmental sources (e.g., cooling tower water). Since the natural mode of transmission appears to be airborne, the greatest potential hazard is the generation of aerosols during the manipulation of cultures or of other concentrations of infectious materials (e.g., infected yolk sacs and tissues.)

Recommended precautions. Biosafety Level 2 practices, containment equipment, and facilities are recommended for all activities involving the use or manipulation of known or potentially infectious clinical materials or cultures and for the housing of infected animals. Primary containment devices and equipment (e.g., biological safety cabinets, centrifuge safety cups) should be used for activities likely to generate potentially infectious aerosols. Vaccines are not available for use in humans.

Agent: ***Mycobacterium leprae***

Inadvertent parenteral human-to-human transmission of leprosy following an accidental needle stick in a surgeon (69) and the use of a presumably contaminated tattoo needle (87) have been reported. There are no cases reported as a result of working in a laboratory with biopsy or other clinical materials of human or animal origin. While naturally occurring leprosy or leprosy-like diseases have been reported in armadillos (120) and in nonhuman primates (35, 76), humans are the only known important reservoir of this disease.

Laboratory hazards. The infectious agent may be present in tissues and exudates from lesions of infected humans and experimentally or naturally infected animals. Direct contact of the skin and mucous membranes with infectious materials and accidental parenteral inoculation are the primary laboratory hazards associated with handling infectious clinical materials.

Recommended precautions. Biosafety Level 2 practices, containment equipment, and facilities are recommended for all activities with known or potentially infectious clinical materials from infected humans and animals. Extraordinary care should be taken to avoid accidental parenteral inoculation with contaminated sharp instruments. Animal Biosafety Level 2 practices and facilities are recommended for animal studies utilizing rodents, armadillos, and nonhuman primates.

Agent: ***Mycobacterium* spp. other than *M. tuberculosis, M. bovis,* or *M. leprae***

Pike reported 40 cases of nonpulmonary "tuberculosis" thought to be related to accidents or incidents in the laboratory or autopsy room (90). Presumably these infections were due to mycobacteria other than *M. tuberculosis* or *M. bovis*. A number of mycobacteria which are ubiquitous in nature are associated with diseases other than tuberculosis or leprosy in humans, domestic animals, and wildlife. Characteristically, these organisms are infectious but not contagious. Clinically, the diseases associated with infections by these "atypical" mycobacteria can be divided into three general categories:

1. **Pulmonary diseases resembling tuberculosis** which may be associated with infection with *M. kansasii, M. avium* complex, and, rarely, with *M. xenopi, M. malmoense, M. asiaticum, M. simiae,* and *M. szulgai*
2. **Lymphadenitis** which may be associated with infection with *M. scrofulaceum, M. avium* complex, and, rarely, with *M. fortuitum and M. kansasii*
3. **Skin ulcers and soft tissue wound infections** which may be associated with infection with

M. ulcerans, M. marinum, M. fortuitum, and *M. chelonei*

Laboratory hazards. The agents may be present in sputa, exudates from lesions, tissues, and environmental samples (e.g., soil and water). Direct contact of skin or mucous membranes with infectious materials, ingestion, and accidental parenteral inoculation are the primary laboratory hazards associated with clinical materials and cultures. Infectious aerosols created during the manipulation of broth cultures or tissue homogenates of these organisms associated with pulmonary disease also pose a potential infection hazard to laboratory personnel.

Recommended precautions. Biosafety Level 2 practices, containment equipment, and facilities are recommended for activities with clinical materials and cultures of *Mycobacterium* spp. other than *M. tuberculosis* or *M. bovis*. Animal Biosafety Level 2 practices and facilities are recommended for animal studies with the mycobacteria other than *M. tuberculosis, M. bovis,* or *M. leprae*.

Agent: ***Mycobacterium tuberculosis, M. bovis***

M. tuberculosis and *M. bovis* infections are a proven hazard to laboratory personnel as well as to others who may be exposed to infectious aerosols in the laboratory (90, 93). The incidence of tuberculosis in laboratory workers working with *M. tuberculosis* is three times higher than that of laboratorians not working with the agent (95). Naturally or experimentally infected nonhuman primates are a proven source of human infection (e.g., the annual tuberculin conversion rate in personnel working with infected nonhuman primates is about 70/10,000 compared with less than 3/10,000 in the general population (62). Experimentally infected guinea pigs or mice do not pose the same problem, since droplet nuclei are not produced by coughing in these species; however, litter from infected animals may become contaminated and serve as a source of infectious aerosols.

Laboratory hazards. Tubercle bacilli may be present in sputum, gastric lavage fluids, cerebrospinal fluid, urine, and lesions from a variety of tissues (3). Exposure to laboratory-generated aerosols is the most important hazard encountered. Tubercle bacilli may survive in heat-fixed smears (1) and may be aerosolized in the preparation of frozen sections and during manipulation of liquid cultures. Because of the low infectious dose of *M. tuberculosis* for humans (i.e., 50% infectious dose equals <10 bacilli) (98, 99) and in some laboratories a high rate of isolation of acid-fast organisms from clinical specimens (>10%) (47), sputa and other clinical specimens from suspected or known cases of tuberculosis must be considered potentially infectious and handled with appropriate precautions.

Recommended precautions. Biosafety Level 2 practices, containment equipment, and facilities (see American Thoracic Society laboratory service levels I and II) (2, 65) are recommended for preparing acid-fast smears and for culturing sputa or other clinical specimens, provided that aerosol-generating manipulations of such specimens are conducted in a Class I or II biological safety cabinet. Liquification and concentration of sputa for acid-fast staining may also be conducted on the open bench at Biosafety Level 2 by first treating the specimen with an equal volume of 5% sodium hypochlorite solution (undiluted household bleach) and waiting 15 min before centrifugation (85, 108).

Biosafety Level 3 practices, containment equipment, and facilities (see American Thoracic Society laboratory service level III) (2, 65) are recommended for activities involving the propagation and manipulation of cultures of *M. tuberculosis* or *M. bovis* and for animal studies utilizing nonhuman primates experimentally or naturally infected with *M. tuberculosis* or *M. bovis*. Animal studies utilizing guinea pigs or mice can be conducted at Animal Biosafety Level 2. Skin testing with purified protein derivative (PPD) of previously skin-tested-negative laboratory personnel can be used as a surveillance procedure. A licensed attenuated live vaccine (BCG) is available but is not routinely used in laboratory personnel.

Agent: ***Neisseria gonorrhoeae***

Four cases of laboratory-associated gonorrhea have been reported in the United States (34, 90).

Laboratory hazards. The agent may be present in conjunctival, urethral, and cervical exudates, synovial fluid, urine, feces, and cerebrospinal fluid. Accidental parenteral inoculation and direct or indirect contact of mucous membranes with infectious clinical materials are the primary laboratory hazards. The importance of aerosols is not determined.

Recommended precautions. Biosafety Level 2 practices, containment equipment, and facilities are

recommended for all activities involving the use or manipulation of clinical materials or cultures. Gloves should be worn when handling infected laboratory animals and when there is the likelihood of direct skin contact with infectious materials. Additional primary containment and personnel precautions, such as those described for Biosafety Level 3, may be indicated for aerosol or droplet production and for activities involving production quantities or concentrations of infectious materials. Vaccines are not available for use in humans.

Agent: ***Neisseria meningitidis***

Meningococcal meningitis is a demonstrated but rare hazard to laboratory workers (4, 92).

Laboratory hazards. The agent may be present in pharyngeal exudates, cerebrospinal fluid, blood, and saliva. Parenteral inoculation, droplet exposure of mucous membranes, and infectious aerosol and ingestion are the primary hazards to laboratory personnel.

Recommended precautions. Biosafety Level 2 practices, containment equipment, and facilities are recommended for all activities utilizing known or potentially infectious body fluids and tissues. Additional primary containment and personnel precautions, such as those described for Biosafety Level 3, may be indicated for activities with high potential for droplet or aerosol production and for activities involving production quantities or concentrations of infectious materials. The use of licensed polysaccharide vaccines (19) should be considered for personnel regularly working with large volumes or high concentrations of infectious materials.

Agent: ***Pseudomonas pseudomallei***

Two laboratory-associated cases of melioidosis are reported, one associated with a massive aerosol and skin contact exposure (48), the second resulting from an aerosol created during the open-flask sonication of a culture presumed to be *Pseudomonas cepacia* (102).

Laboratory hazards. The agent may be present in sputa, blood, wound exudates, and various tissues, depending on site of localization of the infection. Direct contact with cultures and infectious materials from humans, animals, or the environment, ingestion, autoinoculation, and exposure to infectious aerosols and droplets are the primary laboratory hazards. The agent has been demonstrated in blood, sputum, and abscess materials and may be present in soil and water samples from endemic areas.

Recommended precautions. Biosafety Level 2 practices, containment equipment, and facilities are recommended for all activities utilizing known or potentially infectious body fluids and tissues. Gloves should be worn when handling, and during necropsy of, infected animals and when there is the likelihood of direct skin contact with infectious materials. Additional primary containment and personnel precautions, such as those described for Biosafety Level 3, may be indicated for activities with a high potential for aerosol or droplet production and the activities involving production quantities or concentrations of infectious materials.

Agent: ***Salmonella cholera-suis, S. enteritidis*—all serotypes**

Salmonellosis is a documented hazard to laboratory personnel (90). Primary reservoir hosts include a broad spectrum of domestic and wild animals including birds, mammals, and reptiles, all of which may serve as a source of infection to laboratory personnel.

Laboratory hazards. The agent may be present in feces, blood, and urine and in food, feed, and environmental materials. Ingestion or parenteral inoculation are the primary laboratory hazards. The importance of aerosol exposure is not known. Naturally or experimentally infected animals are a potential source of infection for laboratory and animal care personnel and for other animals.

Recommended precautions. Biosafety Level 2 practices, containment equipment, and facilities are recommended for activities with clinical materials known to contain or potentially containing the agents. Animal Biosafety Level 2 practices and facilities are recommended for activities with experimentally or naturally infected animals.

Agent: ***Salmonella typhi***

Typhoid fever is a demonstrated hazard to laboratory personnel (7, 92).

Laboratory hazards. The agent may be present in feces, blood, gallbladder (bile), and urine. Humans are the only known reservoir of infection. Ingestion and parenteral inoculation of the organism represent the primary laboratory hazards. The importance of aerosol exposure is not known.

Recommended precautions. Biosafety Level 2 practices, containment equipment, and facilities are recommended for all activities utilizing known or potentially infectious clinical materials and cultures.

Licensed vaccines, which have been shown to protect 70-90% of recipients, may be a valuable adjunct to good safety practices in personnel regularly working with cultures or clinical materials which may contain *S. typhi* (7).

Agent: ***Shigella* spp.**

Shigellosis is a demonstrated hazard to laboratory personnel, with 49 cases reported in the United States (90). While outbreaks have occurred in captive nonhuman primates, humans are the only significant reservoir of infection. Experimentally infected guinea pigs, other rodents, and nonhuman primates are a proven source of infection.

Laboratory hazards. The agent may be present in feces, and, rarely, in blood of infected humans or animals. Ingestion and parenteral inoculation of the agent are the primary laboratory hazards. The oral 25-50% infectious dose of *S. flexneri* for humans is on the order of 200 organisms (122). The importance of aerosol exposure is not known.

Recommended precautions. Biosafety Level 2 practices, containment equipment, and facilities are recommended for all activities utilizing known or potentially infectious clinical materials or cultures. Animal Biosafety Level 2 facilities and practices are recommended for activities with experimentally or naturally infected animals. Vaccines are not available for use in humans.

Agent: ***Treponema pallidum***

Syphilis is a documented hazard to laboratory personnel who handle or collect clinical material from cutaneous lesions. Pike lists 20 cases of laboratory-associated infection (90). Humans are the only known natural reservoir of the agent.

No cases of laboratory animal-associated infections are reported; however, rabbit-adapted strains of *T. pallidum* (Nichols and possibly others) retain their virulence for humans.

Laboratory hazards. The agent may be present in materials collected from primary and secondary cutaneous lesions and in blood. Accidental parenteral inoculation and contact of mucous membranes or broken skin with infectious clinical materials (and, perhaps, infectious aerosols) are the primary hazards to laboratory personnel.

Recommended precautions. Biosafety Level 2 practices, containment equipment, and facilities are recommended for all activities involving the use or manipulation of blood or lesion materials from humans or infected rabbits. Gloves should be worn when there is a likelihood of direct skin contact with lesion materials. Periodic serological monitoring should be considered in personnel regularly working with infectious materials. Vaccines are not available for use in humans.

Agent: **Vibrionic enteritis (*Campylobacter fetus* subsp. *jejuni, Vibrio cholerae, V. parahaemolyticus*)**

Vibrionic enteritis due to *C. fetus jejuni, V. cholerae,* or *V. parahaemolyticus* is a documented but rare cause of laboratory-associated illnesses (92). Naturally and experimentally infected animals are a potential source of infection (94).

Laboratory hazards. All pathogenic vibrios may occur in feces. *C. fetus* may also be present in blood, exudates from abscesses, tissue, and sputa. Ingestion of *V. cholerae* and ingestion or parenteral inoculation of other vibrios constitute the primary laboratory hazard. The human oral infecting dose of *V. cholerae* in healthy non-achlorhydric individuals is of the order of 10^8 organisms (94). The importance of aerosol exposure is not known. The risk of infection following oral exposure may be increased in achlorhydric individuals.

Recommended precautions. Biosafety Level 2 practices, containment equipment, and facilities are recommended for activities with cultures or potentially infectious clinical materials. Animal Biosafety Level 2 practices and facilities are recommended for activities with naturally or experimentally infected animals. Although vaccines have been shown to provide partial protection of short duration (3-6 months) to nonimmune individuals in highly endemic areas (7), the routine use of cholera vaccine in laboratory staff is not recommended.

Agent: ***Yersinia pestis***

Plague is a proven but rare laboratory hazard. Four cases have been reported in the United States (11, 90).

Laboratory hazards. The agent may be present in bubo fluid, blood, sputum, cerebrospinal fluid, feces, and urine from humans, depending on the clinical form and stage of the disease. Direct contact with cultures and infectious materials from humans or rodents, infectious aerosols or droplets generated during the manipulation of cultures and infected tissues and in the necropsy of rodents, accidental autoinoculation, ingestion, and bites by infected fleas collected from rodents are the primary hazards to laboratory personnel.

Recommended precautions. Biosafety Level 2 practices, containment equipment, and facilities are recommended for all activities involving the handling of potentially infectious clinical materials and cultures. Special care should be taken to avoid the generation of aerosols of infectious materials and during the necropsy of naturally or experimentally infected rodents. Gloves should be worn when handling field-collected or infected laboratory rodents and when there is the likelihood of direct skin contact with infectious materials. Necropsy of rodents is ideally conducted in a biological safety cabinet. Although field trials have not been conducted to determine the efficacy of licensed inactivated vaccines, experience with these products has been favorable (14). Immunization is recommended for personnel working regularly with cultures of *Y. pestis* or infected rodents (26).

Additional primary containment and personnel precautions, such as those described for Biosafety Level 3, are recommended for activities with high potential for droplet or aerosol production, for work with antibiotic-resistant strains, and for activities involving production quantities or concentrations of infectious materials.

Rickettsial Agents

Agent: ***Coxiella burnetii***

Pike's summary indicates that Q fever is the second most commonly reported laboratory-associated infection, with outbreaks involving 15 or more persons recorded in several institutions (90). A broad range of domestic and wild mammals are natural hosts for Q fever and may serve as potential sources of infection for laboratory and animal care personnel. Exposure to naturally infected and often asymptomatic sheep and to their birth products is a documented hazard to personnel (20, 109). The agent is remarkably resistant to drying and is stable under a variety of environmental conditions (121).

Laboratory hazards. The agent may be present in infected arthropods and in the blood, urine, feces, milk, or tissues of infected animal or human hosts. The placenta of infected sheep may contain as many as 10^9 organisms per gram of tissue, and milk may contain 10^5 organisms per gram. Parenteral inoculation and exposure to infectious aerosols and droplets are the most likely sources of infection for laboratory and animal care personnel. The estimated human 25-50% infectious dose (inhalation) for Q fever is 10 organisms (122).

Recommended precautions. Biosafety Level 2 practices, containment equipment, and facilities are recommended for nonpropagative laboratory procedures, including serological examinations and staining of impression smears. Biosafety Level 3 practices and facilities are recommended for activities involving the inoculation, incubation, and harvesting of embryonated eggs or tissue cultures, the necropsy of infected animals, and the manipulation of infected tissues. Since infected guinea pigs and other rodents may shed the organisms in urine or feces (90), experimentally infected rodents should be maintained under Animal Biosafety Level 3. Recommended precautions for facilities using sheep as experimental animals are described by Spinelli (109) and by Bernard (6). An Investigational New Phase-1 Q fever vaccine (IND) is available from the U.S. Army Medical Research Institute for Infectious Diseases, Fort Detrick, Maryland. The use of this vaccine should be limited to those at high risk of exposure who have no demonstrated sensitivity to Q fever antigen.

Agent: ***Rickettsia akari, Rochalimaea quintana,* and *Rochalimaea vinsonii***

Based on the experience of laboratories actively working with *Rickettsia akari*, it is likely that the five cases of rickettsialpox recorded by Pike (90) were associated with exposure to bites of infected mites rather than aerosol or contact exposure to infected tissues. There are no recorded cases of laboratory-associated infections with trench fever (*Rochalimaea quintana*) or vole rickettsia (*Rochalimaea vinsonii*).

Laboratory hazards. The agent of rickettsialpox may be present in blood and other tissues of infected house mice or humans and in the mite vector *Liponyssoides sanguineus*. Exposure to naturally or experimentally infected mites and accidental parenteral inoculation are the most likely sources of human infection with rickettsialpox. The agent of trench fever may be present in the blood and tissues of infected humans and in the body fluids and feces of infected human body lice (*Pediculus h. humanus*).

Recommended precautions. Biosafety Level 2 practices, containment equipment, and facilities are recommended for propagation and animal studies with *Rickettsia akari, Rochalimaea vinsonii,* and *Rochalimaea quintana*. Appropriate precautions should be taken to avoid exposure of personnel to infected mites that are maintained in the laboratory or that may be present on naturally infected house mice.

Agent: ***Rickettsia prowazekii, Rickettsia typhi (R. mooseri), Rickettsia tsutsugamushi, Rickettsia canada,* and Spotted Fever Group agents of human disease other than *Rickettsia rickettsii* and *Rickettsia akari***

Pike reported 57 cases of laboratory-associated typhus (type not specified), 56 cases of epidemic typhus with three deaths, and 68 cases of murine typhus (90). More recently, three cases of murine typhus were reported from a research facility (18). Two of these three cases were associated with work with infectious materials on the open bench; the third case resulted from an accidental parenteral inoculation. These three cases represented an attack rate of 20% in personnel working with infectious materials.

Laboratory hazards. Accidental parenteral inoculation and exposure to infectious aerosols are the most likely sources of laboratory-associated infections. Naturally or experimentally infected lice, fleas, and flying squirrels (*Glaucomys* spp.) (9) may also be a direct source of infection to laboratory personnel. The organisms are relatively unstable under ambient environmental conditions.

Recommended precautions. Biosafety Level 2 practices, containment equipment, and facilities are recommended for nonpropagative laboratory procedures, including serological and fluorescent-antibody procedures, and for the staining of impression smears. Biosafety Level 3 practices and facilities are recommended for all other manipulations of known or potentially infectious materials, including necropsy of experimentally infected animals and trituration of their tissues and inoculation, incubation, and harvesting of embryonated eggs or tissue cultures. Animal Biosafety Level 2 practices and facilities are recommended for activities with infected mammals other than flying squirrels or arthropods. Vaccines are not currently available for use in humans. Because the mode of transmission of *R. prowazekii* from flying squirrels to humans is not defined, Animal Biosafety Level 3 practices and facilities are recommended for animal studies with flying squirrels naturally or experimentally infected with *R. prowazekii*.

Agent: ***Rickettsia rickettsii***

Rocky Mountain spotted fever is a documented hazard to laboratory personnel. Pike (90) reported 63 laboratory-associated cases, 11 of which were fatal. Oster (86) reported nine cases occurring over a 6-year period in one laboratory which were believed to have been acquired as a result of exposure to infectious aerosols.

Laboratory hazards. Accidental parenteral inoculation and exposure to infectious aerosols are the most likely sources of laboratory-associated infection (57). Successful aerosol transmission has been experimentally documented in nonhuman primates (101). Naturally and experimentally infected mammals, their ectoparasites, and their infected tissues are sources of human infection. The organism is relatively unstable under ambient environmental conditions.

Recommended precautions. Biosafety Level 2 practices, containment equipment, and facilities are recommended for all nonpropagative laboratory procedures, including serological and fluorescent-antibody tests, and staining of impression smears. Biosafety Level 3 practices and facilities are recommended for all other manipulations of known or potentially infectious materials, including necropsy of experimentally infected animals and trituration of their tissues and inoculation, incubation, and harvesting of embryonated eggs or tissue cultures. Animal Biosafety Level 2 practices and facilities are recommended for holding of experimentally infected rodents; however, necropsy and any subsequent manipulation of tissues from infected animals should be conducted at Biosafety Level 3.

Because of the proven value of antibiotic therapy in the early stages of infection, it is essential that laboratories working with *R. rickettsii* have an effective system for reporting febrile illnesses in laboratory personnel, medical evaluation of potential cases, and, when indicated, institution of appropriate antibiotic therapy. Vaccines are not currently available for use in humans (see Appendix A.3).

Viral Agents

Agent: **Hepatitis A virus**

Laboratory-associated infections with hepatitis A virus do not appear to be an important occupational risk among laboratory personnel. However, the disease is a documented hazard in animal handlers and others working with chimpanzees which are naturally or experimentally infected (92).

Laboratory hazards. The agent may be present in feces of infected humans and chimpanzees. Ingestion of feces, stool suspensions, and other contaminated materials is the primary hazard to laboratory personnel. The importance of aerosol exposure has not been demonstrated. Attenuated or avirulent strains have not been fully defined but appear to result from serial passage in tissue culture.

Recommended precautions. Biosafety Level 2 practices, containment equipment, and facilities are recommended for activities with known or potentially infected feces from humans or chimpanzees. Animal Biosafety Level 2 practices and facilities are recommended for activities using naturally or experimentally infected chimpanzees. Animal care personnel should wear gloves and take other appropriate precautions to avoid possible fecal-oral exposure. Vaccines are not available for use in humans, but are in the developmental stages.

Agent: **Hepatitis B, hepatitis non-A non-B**

Pike concluded that hepatitis B is currently the most frequently occurring laboratory-associated infection (90). The incidence in some categories of laboratory workers is seven times greater than that of the general population (104). Epidemiological evidence indicates that hepatitis non-A non-B is a bloodborne disease similar to hepatitis B.

Laboratory hazards. The agent of hepatitis B may be present in blood and blood products of human origin and in urine, semen, cerebrospinal fluid, and saliva. Parenteral inoculation, droplet exposure of mucous membranes, and contact exposure of broken skin are the primary laboratory hazards. The virus may be stable in dried blood or blood components for several days. Attenuated or avirulent strains are not defined.

Recommended precautions. Biosafety Level 2 practices, containment equipment, and facilities are recommended for all activities utilizing known or potentially infectious body fluids and tissues. Additional primary containment and personnel precautions, such as those described for Biosafety Level 3, may be indicated for activities with high potential for droplet or aerosol production and for activities involving production quantities or concentrations of infectious materials. Animal Biosafety Level 2 practices, containment equipment, and facilities are recommended for activities utilizing naturally or experimentally infected chimpanzees or other nonhuman primates. Gloves should be worn when working with infected animals and when there is the likelihood of skin contact with infectious materials. A licensed inactivated vaccine is available and is recommended for laboratory personnel, who are at substantially greater risk of acquiring infection than is the general population (27).

Agent: ***Herpesvirus simiae*** **(B-virus)**

Although B-virus presents a potential hazard to laboratory personnel working with the agent, laboratory-associated human infections with B-virus have, with rare exceptions, been limited to personnel having direct contact with living Old World monkeys (29, 56, 89). Exposure to in vitro monkey tissues (i.e., primary rhesus monkey kidney) has been associated with a single documented case (29).

B-virus is an indigenous chronic and/or recurrent infection of macaques and possibly other Old World monkeys and is a frequent enzootic infection of captive *Macaca mulatta*.

Laboratory personnel handling Old World monkeys run the risk of acquiring B-virus from a bite or contamination of broken skin or mucous membranes by an infected monkey. Fifteen fatal cases of human infections with B-virus have been reported (29).

Laboratory hazards. The agent may be present in oral secretions, thoracic and abdominal viscera, and central nervous system tissues of naturally infected macaques. Bites from monkeys with oral herpes lesions are the greatest hazard to laboratory and animal care personnel. Exposures of broken skin or mucous membranes to oral secretions or to infectious culture fluids are also potential hazards. The importance of aerosol exposure is not known. Attenuated or avirulent strains have not been defined.

Recommended precautions. Biosafety Level 2 practices, containment equipment, and facilities are recommended for all activities involving the use or manipulation of tissues, body fluids, and primary tissue culture materials from macaques. Additional containment and personnel precautions, such as those recommended for Biosafety Level 3, are recommended for activities involving the use or manipulation of any material known to contain *H. simiae.*

Biosafety Level 4 practices, containment equipment, and facilities are recommended for activities involving the propagation of *H. simiae,* manipulations of production quantities or concentrations of *H. simiae,* and housing vertebrate animals with proven natural or induced infection with the agent.

The wearing of gloves, masks, and laboratory coats is recommended for all personnel working with nonhuman primates—especially macaques and other Old World species—and for all persons entering animal rooms where nonhuman primates are housed. Vaccines are not available for use in humans.

Agent: **Herpesviruses**

The herpesviruses are ubiquitous human pathogens and are commonly present in a variety of clinical materials submitted for virus isolation. While these viruses are not demonstrated causes of laboratory-associated infections, they are primary as well as opportunistic pathogens, especially in immunocompromised hosts. Nonpolio enteroviruses, adenoviruses, and cytomegalovirus pose similar low potential infection risks to laboratory personnel. Although this diverse group of indigenous viral agents does not meet the criteria for inclusion in agent-specific summary statements (i.e., demonstrated or high potential hazard for laboratory-associated infection; grave consequences should infection occur), the frequency of their presence in clinical materials and their common use in research warrants their inclusion in this publication.

Laboratory hazards. Clinical materials and isolates of herpesviruses, nonpolio enteroviruses, and other indigenous pathogens may pose a risk of infection following ingestion, accidental parenteral inoculation, droplet exposure of the mucous membranes of the eyes, nose, or mouth, or inhalation of concentrated aerosolized materials.

Recommended precautions. Biosafety Level 2 practices, containment equipment, and facilities are recommended for activities utilizing known or potentially infectious clinical materials or cultures of indigenous viral agents which are associated with or identified as a primary pathogen of human disease. Although there is no definitive evidence that infectious aerosols are a significant source of laboratory-associated infections, it is prudent to avoid the generation of aerosols during the handling of clinical materials and isolates or during the necropsy of animals. Primary containment devices (e.g., biological safety cabinets) constitute the basic barrier protecting personnel from exposure to infectious aerosols.

Agent: **Influenza virus**

Laboratory-associated infections with influenza are not normally documented in the literature but are known to occur by informal accounts and published reports, particularly when new strains showing antigenic drift or shift are introduced into a laboratory for diagnostic or research purposes (36).

Laboratory animal-associated infections are not reported; however there is a high possibility of human infection from infected ferrets and vice versa.

Laboratory hazards. The agent may be present in respiratory tissues or secretions of man or most infected animals and in the cloaca of many infected avian species. The virus may be disseminated in multiple organs in some infected animal species.

Inhalation of virus from aerosols generated by aspirating, dispensing, or mixing virus-infected samples or by infected animals is the primary laboratory hazard. Genetic manipulation of virus has unknown potential for altering host range and pathogenicity or for introducing into man transmissible viruses with novel antigenic composition.

Recommended precautions. Biosafety Level 2 practices, containment equipment, and facilities are

recommended when receiving and inoculating routine laboratory diagnostic specimens. Autopsy material should be handled in a biological safety cabinet using Biosafety Level 2 procedures.

Activities utilizing noncontemporary virus strains. Biosafety considerations should take into account the available information about infectiousness of the strains being used and the potential for harm to the individual or society in the event that laboratory-acquired infection and subsequent transmission occur. Research or production activities utilizing contemporary strains may be safely performed using Biosafety Level 2 containment practices. Susceptibility to infection with older noncontemporary human strains, with recombinants, or with animal isolates warrants the use of Biosafety Level 2 containment procedures. Current experience suggests, however, that there is no evidence for laboratory-acquired infection with reference strains A/PR/8/34 and A/WS/33 or their commonly used neurotropic variants.

Agent: **Lymphocytic choriomeningitis (LCM) virus**

Laboratory-associated infections with lymphocytic choriomeningitis virus are well documented in facilities where infections occur in laboratory rodents, especially mice and hamsters (8, 90). Tissue cultures which have inadvertently become infected represent a potential source of infection and dissemination of the agent. Natural infections are occasionally found in nonhuman primates, swine, and dogs.

Laboratory hazards. The agent may be present in blood, cerebrospinal fluid, urine, secretions of the nasopharynx, feces, and tissues of infected humans and other animal hosts. Parenteral inoculation, inhalation, and contamination of mucous membranes or broken skin with infectious tissues or fluids from infected animals are common hazards. Aerosol transmission is well documented (8).

Recommended precautions. Biosafety Level 2 practices, containment equipment, and facilities are recommended for all activities utilizing known or potentially infectious body fluids or tissues and for tissue culture passage of mouse brain-passaged strains. All manipulations of known or potentially infectious passage and clinical materials should be conducted in a biological safety cabinet. Additional primary containment and personnel precautions such as those described for Biosafety Level 3 may be indicated for activities with high potential for aerosol production and for activities involving production quantities or concentrations of infectious materials. Animal Biosafety Level 2 practices and facilities are recommended for studies in adult mice with mouse brain-passaged strains. Animal Biosafety Level 3 practices and facilities are recommended for work with infected hamsters. Vaccines are not available for use in humans.

Agent: **Poliovirus**

Laboratory-associated infections with polioviruses are uncommon and are generally limited to unvaccinated laboratory persons working directly with the agent. Twelve cases have been reported in the world literature (90).

Laboratory animal-associated infections have not been reported (73); however, naturally or experimentally infected nonhuman primates could provide a source of infection to exposed unvaccinated persons.

Laboratory hazards. The agent may be found in the feces and in throat secretions. Ingestion and parenteral inoculation of infectious tissues or fluids by unimmunized personnel are the primary hazards to laboratory personnel. The importance of aerosol exposure is not known. Laboratory exposures pose negligible risk to appropriately immunized persons.

Recommended precautions. Biosafety Level 2 practices, containment equipment, and facilities are recommended for all activities utilizing known or potentially infectious culture fluids and specimen materials. All laboratory personnel working directly with the agent must have documented polio vaccination or demonstrated serologic evidence of immunity to all three poliovirus types (25).

Agent: **Poxviruses**

Sporadic cases of laboratory-associated poxvirus infections have been reported. Pike lists 24 cases of yaba and tanapox virus infection and 18 vaccinia and smallpox infections (90). Epidemiological evidence suggests that transmission of monkeypox virus from nonhuman primates or rodents to humans may have occurred in nature but not in the laboratory setting. Naturally or experimentally infected laboratory animals are a potential source of infection to exposed unvaccinated laboratory personnel.

Laboratory hazards. The agents may be present in lesion fluids or crusts, respiratory secretions, or tissues of infected hosts. Ingestion, parenteral inoculation, and droplet or aerosol exposure of mucous membranes or broken skin to infectious fluids or tissues are the primary hazards to laboratory and animal care personnel. Some poxviruses are stable at ambient temperature when dried and may be transmitted by fomites.

Recommended precautions. The possession and use of variola viruses are restricted to the World Health Organization Collaborating Center for Smallpox and Other Poxvirus Infections located at the Centers for Disease Control, Atlanta, Georgia. Biosafety Level 2 practices, containment equipment, and facilities are recommended for all activities involving the use or manipulation of poxviruses other than variola that pose an infection hazard to humans. All persons working in or entering laboratory or animal care areas where activities with vaccinia, monkeypox, or cowpox viruses are being conducted should have documented evidence of satisfactory vaccination within the preceding 3 years (23). Activities with vaccinia, cowpox, or monkeypox viruses in quantities or concentrations greater than those present in diagnostic cultures may also be conducted by immunized personnel at Biosafety Level 2, provided that all manipulations of viable materials are conducted in Class I or II biological safety cabinets or other primary containment equipment.

Agent: **Rabies virus**

Laboratory-associated rabies infections are extremely rare. Two have been documented. Both resulted from presumed exposure to high-titered infectious aerosols generated in a vaccine production facility (129) and a research facility (17). Naturally or experimentally infected animals, their tissues, and their excretions are a potential source of exposure to laboratory and animal care personnel.

Laboratory hazards. The agent may be present in all tissues of infected animals. Highest titers are demonstrated in central nervous system tissue, salivary glands, and saliva. Accidental parenteral inoculation, cuts, or sticks with contaminated laboratory equipment, bites by infected animals, and exposure of mucous membranes or broken skin to infectious droplets of tissue or fluids are the most likely source of exposure for laboratory and animal care personnel. Infectious aerosols have not been a demonstrated hazard to personnel working with clinical materials and conducting diagnostic examinations. Fixed and attenuated strains of virus are presumed to be less hazardous, but the only two recorded cases of laboratory-associated rabies resulted from exposure to a fixed challenge virus standard (CVS) derived from SAD (Street Alabama Dufferin) strain (17, 129).

Recommended precautions. Biosafety Level 2 practices, containment equipment, and facilities are recommended for all activities utilizing known or potentially infectious materials. Preexposure immunization is recommended for all individuals working with rabies virus or infected animals or engaged in diagnostic, production, or research activities with rabies virus. Preexposure immunization is also recommended for all individuals entering or working in the same room where rabies virus or infected animals are used. When it is not feasible to open the skull or remove the brain within a biological safety cabinet, it is pertinent to wear heavy protective gloves to avoid cuts or sticks from cutting instruments or bone fragments, and to wear a face shield to protect the mucous membranes of the eyes, nose, and mouth from exposure to infectious droplets or tissue fragments. If a Stryker saw is used to open the skull, avoid striking the brain with the blade of the saw. Additional primary containment and personnel precautions, such as those described for Biosafety Level 3, may be indicated for activities with a high potential for droplet or aerosol production and for activities involving production quantities or concentrations of infectious materials.

Agents: **Transmissible spongiform encephalopathies (Creutzfeldt-Jakob and kuru agents)**

Laboratory-associated infections with the transmissible spongiform encephalopathies have not been documented. The consequences of infection are grave, however, and there is evidence that Creutzfeldt-Jakob disease has been transmitted to patients by corneal transplant and by contaminated electroencephalographic electrodes. There is no known nonhuman reservoir for Creutzfeldt-Jakob disease or kuru. Nonhuman primates and other laboratory animals have been infected by inoculation, but there is no evidence of secondary transmission.

Laboratory hazards. High titers of a transmissible agent have been demonstrated in the brain and spinal cord of persons with kuru. In persons with Creutzfeldt-Jakob disease, a transmissible agent has been demonstrated in the brain, spleen, liver, lymph nodes, lungs, spinal cord, kidneys, cornea, and lens. Accidental parenteral inoculation, especially with neural tissues, and including Formalin-fixed specimens, is extremely hazardous. Although nonneural tissues are less often infective, all tissues of humans and animals infected with these agents should be considered potentially hazardous. The risk of infection from aerosols and droplets and from exposure to intact skin and gastric and mucous membranes is not known; however, there is no evidence of contact or aerosol transmission. These agents are characterized by extreme resistance to conventional inactivation procedures, including irradiation, boiling, and chemicals (Formalin, betapropiolactone, alcohols).

Recommended precautions. Biosafety Level 2 practices, containment equipment, and facilities are recommended for all activities utilizing known or potentially infectious tissues and fluids from naturally infected humans and from experimentally infected animals. Extreme care must be taken to avoid accidental autoinoculation or other traumatic parenteral inoculations of infectious tissues and fluids (46). Although there is no evidence to suggest that aerosol transmission occurs in the natural disease, it is prudent to avoid the generation of aerosols or droplets during the manipulation of tissues and fluids and during the necropsy of experimental animals. It is further recommended that gloves should be worn for activities which provide the opportunity for skin contact with infectious tissues and fluids. Vaccines are not available for use in humans.

Agent: **Vesicular stomatitis virus (VSV)**

Forty-six laboratory-associated infections with indigenous strains of VSV have been reported (112). Laboratory activities with indigenous strains of VSV present two different levels of risk to laboratory personnel and are related, at least in part, to the passage history of the strains utilized. Activities utilizing infected livestock, their infected tissues, and virulent isolates from these sources are a demonstrated hazard to laboratory and animal care personnel (52, 88). Seroconversion and clinical illness rates in personnel working with these materials are high (88). Similar risks may be associated with exotic strains such as Piry (112).

In contrast, anecdotal information indicates that activities with less virulent laboratory-adapted strains (e.g., VSV-Indiana [San Juan and Glasgow]) are rarely associated with seroconversion or illness. Such strains are commonly used by molecular biologists, often in large volumes and high concentrations, under conditions of minimal or no primary containment. Experimentally infected mice have not been a documented source of human infection.

Laboratory hazards. The agent may be present in vesicular fluid, tissues, and blood of infected animals and in blood and throat secretions of infected humans. Exposure to infectious aerosols and infected droplets, direct skin and mucous membrane contact with infectious tissues and fluids, and accidental autoinoculation are the primary laboratory hazards associated with virulent isolates. Accidental parenteral inoculation and exposure to infectious aerosols represent potential risks to personnel working with less virulent laboratory-adapted strains.

Recommended precautions. Biosafety Level 3 practices, containment equipment, and facilities are recommended for activities involving the use or manipulation of infected tissues and of virulent isolates from naturally or experimentally infected livestock. Gloves and respiratory protection are recommended for the necropsy and handling of infected animals. Biosafety Level 2 practices and facilities are recommended for activities utilizing laboratory-adapted strains of demonstrated low virulence. Vaccines are not available for use in humans.

ARBOVIRUSES

Arboviruses Assigned to Biosafety Level 2

The American Committee on Arthropod-Borne Viruses (ACAV) registered 424 arboviruses as of December 31, 1979. The ACAV's Subcommittee on Arbovirus Laboratory Safety (SALS) has categorized each of these 424 agents into one of four recommended levels of practice and containment which parallel the recommended practices, safety equipment, and facilities described in this publication as Biosafety Levels 1 - 4 (112). It is the intent of SALS to periodically update the 1980 publication by providing a supplemental listing and recommended levels of practice and containment for arboviruses registered since 1979. SALS categorizations were based on risk assessments from information provided by a worldwide survey of 585 laboratories working with arboviruses. SALS recommended that work with the majority of these agents should be conducted at the equivalent of Biosafety Level 2. These viruses are listed alphabetically on pages 124-126 and include the following agents which are the reported cause of laboratory-associated infections (53, 90, 112).

Virus	Cases (SALS)	
Vesicular stomatitis	46	
Colorado tick fever	16	
Dengue	11	
Pichinde	17	
Western equine encephalomyelitis	7	(2 deaths)
Rio Bravo	7	
Kunjin	6	
Catu	6	
Caraparu	5	
Ross River	5	
Bunyamwera	4	
Eastern equine encephalomyelitis	4	
Zika	4	
Apeu	2	
Marituba	2	
Tacaribe	2	
Muructucu	1	
O'nyong nyong	1	
Modoc	1	
Oriboca	1	
Ossa	1	
Keystone	1	
Bebaru	1	
Bluetongue	1	

The list of arboviruses in Biosafety Level 2 includes yellow fever virus (17D strain) and Venezuelan equine encephalomyelitis (VEE) virus (TC83 strain), provided that personnel working with these vaccine strains are immunized.

The results of the SALS survey clearly indicate that the suspected source of the laboratory-associated infections listed above was other than exposure to infectious aerosols. Recommendations that work with these 334 arboviruses should be conducted at Biosafety Level 2 were based on the existence of adequate historical laboratory experience to assess risks for the virus which indicated that: (a) no overt laboratory-associated infections are reported; or (b) infections resulted from exposures other than to infectious aerosols; or (c) if aerosol exposures are documented they represent an uncommon route of exposure.

Laboratory hazards. Agents listed in this group may be present in blood, cerebrospinal fluid, central nervous system and other tissues, and infected arthropods, depending on the agent and the stage of infection. While the primary laboratory hazards are accidental parenteral inoculation, contact of the virus with broken skin or mucous membranes, and bites by infected laboratory rodents or arthropods, infectious aerosols may also be a potential source of infection.

Arboviruses Assigned to Biosafety Level 2

Abu Hammad	Bertioga
Acado	Bimiti
Acara	Birao
Aguacate	Bluetongue (indigenous)
Alfuy	Boraceia
Almpiwar	Botambi
Amapari	Boteke
Anhanga	Bouboui
Anhembi	Bujaru
Anopheles A	Bunyamwera
Anopheles B	Burg el Arab
Apeu	Bushbush
Apoi	Bussuquara
Aride	Buttonwillow
Arkonam	Bwamba
Aruac	Cacao
Arumowot	Cache Valley
Aura	Caimito
Avalon	California encephalitis
Bagaza	Calovo
Bahig	Candiru
Bakau	Cape Wrath
Baku	Capim
Bandia	Caraparu
Bangoran	Carey Island
Bangui	Catu
Banzi	Chaco
Barur	Chagres
Batai	Chandipura
Batu	Changuinola
Bauline	Charleville
Bebaru	
Belmont	

Chenuda
Chilibre
Chobar
Gorge
Clo Mor
Colorado Tick
Fever
Corriparta
Cotia
Cowbone Ridge
D'Aguilar
Dakar Bat
Dengue-1
Dengue-2
Dengue-3
Dengue-4
Dera Ghazi Khan
Eastern equine
encephalomyeli-
tis
Edge Hill
Entebbe Bat
Epizootic
hemorrhagic
disease
Eubenangee
Eyach
Flanders
Fort Morgan
Frijoles
Gamboa
Gomokà
Gossas
Grand Arbaud
Great Island
Guajara
Guama
Guaroa
Gumbo limbo
Hart Park
Hazara
Huacho
Hughes
Icoaraci
Ieri
Iheus
Ilesha
Ingwavuma
Inkoo
Ippy
Irituia
Isfahan
Itaporanga
Itaqui
Jamestown
Canyon
Japanaut
Jerry Slough
Johnston Atoll
Joinjakaka
Juan Diaz
Jugra
Jurona
Jutiapa
Kadam
Kaeng khoi
Kaikalur
Kaisodi
Kamese
Kammamavanpet-
tai
Kannamangalam
Kao shuan
Karimabad
Karshi
Kasba
Kemervo
Kern Canyon
Ketapang
Keterah
Keuraliba
Keystone
Klamath
Kokobera
Kolongo
Koongol
Kowanyama
Kunjim
Kununurra
Kwatta
La Crosse
Lagos bat
La Joya
Landjia
Langat
Lanjan
Latino
Lebombo
Le Dantec
Lipovnik
Lokern
Lone Star
Lukuni
M'poko
Madrid
Maguari
Mahogany
Hammock
Main Drain
Malakal
Manawa
Manzanilla
Mapputa
Maprik
Marco
Marituba
Matariya
Matruh
Matucare
Melao
Mermet
Minatitlan
Minnal
Mirim
Mitchell River
Modoc
Moju
Mono Lake
Mont. myotis leuk.
Moriche
Mossuril
Mount Elgon bat
Murutucu
Navarro
Nepuyo
Ngaingan
Nique
Nkolbisson
Nola
Ntaya
Nugget
Nyamanini
Nyando
O'nyong-nyong
Okhotskiy
Okola
Olifantsvlei
Oriboca
Ossa
Pacora
Pacui
Pahayokee
Palyam
Parana
Pata
Pathum Thani
Patois
Phnom-Penh bat
Pichinde
Pixuna
Pongola
Pretoria
Puchong
Punta Salinas
Punta Toro
Qalyub
Quaranfil
Restan
Rio Bravo
Rio Grande
Ross River
Royal Farm
Sabo
Saboya
Saint Floris
Sakhalin
Salehabad
San Angelo
Sandfly F. (Naples)
Sandfly F. (Sicilian)
Sandjimba
Sathuperi
Sawgrass
Sebokele
Seletar
Sembalam
Shamonda
Shark River
Shuni
Silverwater
Simbu
Simian hemorrhagic
fever
Sindbis
Sixgun City
Snowshoe hare
Sokuluk
Soldado
Sororoca
Stratford

Sunday Canyon	Thottapalayam	Urucuri	Whataro
Tacaiuma	Timbo	Usutu	Witwatersrand
Tacaribe	Toure	Uukuniemi	Wongal
Taggert	Tribec	Vellore	Wongorr
Tahyna	Triniti	Venezuelan equine encephalomyelitis (TC-83)	Wyeomyia
Tamiami	Trivittatus	Venkatapuram	Yaquina Head
Tanga	Trubanaman	Vesicular stomatitis (see p. 123)	Yata
Tanjong Rabok	Tsuruse	Wad Medani	Yellow fever (17D)
Tataguine	Turlock	Wallal	Yogue
Tembe	Tyuleniy	Wanowrie	Zaliv Terpeniya
Tembusu	Uganda S	Warrego	Zegla
Tensaw	Umatilla	Western equine encephalomyelitis	Zika
Tete	Umbre		Zingilamo
Tettnang	Una		Zirqa
Thimiri	Upolu		

Recommended precautions. Biosafety Level 2 practices, safety equipment, and facilities are recommended for activities with potentially infectious clinical materials and arthropods and for manipulations of infected tissue cultures, embryonated eggs, and rodents. Infection of newly hatched chickens with eastern and western equine encephalomyelitis viruses is especially hazardous and should be undertaken under Biosafety Level 3 conditions by immunized personnel. Investigational vaccines (IND) against eastern equine encephalomyelitis and western equine encephalomyelitis viruses are available through the Centers for Disease Control and the U.S. Army Medical Research Institute for Infectious Diseases, respectively. The use of these vaccines is recommended for personnel who work directly and regularly with these two agents in the laboratory. Western equine encephalomyelitis immune globulin (human) is also available from the Centers for Disease Control. The efficacy of this product has not been established.

Arboviruses and Arenaviruses Assigned to Biosafety Level 3

SALS has recommended that work with the arboviruses included in the alphabetical listing on page 127 should be conducted at the equivalent of Biosafety Level 3 practices, safety equipment, and facilities. These recommendations are based on one of the following criteria: overt laboratory-associated infections with these agents have occurred by aerosol route if protective vaccines are not used or are unavailable, or laboratory experience with the agent is inadequate to assess risk and the natural disease in humans is potentially severe or life threatening or causes residual damage. Hantaan virus, which was not included in the SALS publication, has been placed at Biosafety Level 3 based on documented laboratory-associated infections. Rift Valley fever virus, which was classified by SALS at Containment Level 3 (i.e., HEPA filtration required for all air exhausted from the laboratory), was placed in Biosafety Level 3 provided that all personnel entering the laboratory or animal care area where work with this virus is being conducted are vaccinated. Laboratory or laboratory animal-associated infections have been reported with the following agents (53, 90, 112, 124).

Virus	Cases (SALS)	
Venezuelan equine encephalitis	150	(1 death)
Rift Valley fever	47	(1 death)
Chikungunya	39	
Yellow fever	38	(8 deaths)
Japanese encephalitis	22	
Louping ill	22	
West Nile	18	
Lymphocytic choriomeningitis	15	
Orungo	13	
Piry	13	
Wesselsbron	13	
Mucambo	10	
Oropouche	7	
Germiston	6	

Bhanja	6	
Hantaan (Korean hemorrhagic fever)	6	
Mayaro	5	
Spondweni	4	
St. Louis encephalitis	4	
Murray Valley encephalitis	3	
Semliki Forest	3	(1 death)
Powassan	2	
Dugbe	2	
Issyk-kul	1	
Koutango	1	

Large quantities and high concentrations of Semliki Forest virus are commonly used or manipulated by molecular biologists under conditions of moderate or low containment. Although antibodies have been demonstrated in individuals working with this virus, the first overt (and fatal) laboratory-associated infection with this virus was reported in 1979 (126). Because this infection may have been influenced by a compromised host, an unusual route of exposure or high dosage, or a mutated strain of the virus, this case and its outcome may not be typical. Since exposure to an infectious aerosol was not indicated as the probable mode of transmission in this case, it is suggested that most activities with Semliki Forest disease virus can be safely conducted at Biosafety Level 2.

Some viruses (e.g., Ibaraki, Israel turkey meningoencephalitis) are listed by SALS in Level 3, not because they pose a threat to human health, but because they are exotic diseases of domestic livestock or poultry.

Laboratory hazards. The agents listed in this group may be present in blood, cerebrospinal fluid, urine, and exudates, depending on the specific agent and stage of disease. The primary laboratory hazards are exposure to aerosols of infectious solutions and animal bedding, accidental parenteral inoculation, and broken skin contact. Some of these agents (e.g., Venezuelan equine encephalitis [VEE]) may be relatively stable in dried blood or exudates. Attenuated strains are identified in a number of the agents listed (e.g., yellow fever 17D strain and VEE TC83 strain).

Recommended precautions. Biosafety Level 3 practices, containment equipment, and facilities are recommended for activities using potentially infectious clinical materials and infected tissue cultures, animals, or arthropods.

A licensed attenuated live virus is available for immunization against yellow fever and is recommended for all personnel who work with this agent or with infected animals and for those who enter rooms where the agents or infected animals are present. An investigational vaccine (IND) is available for immunization against VEE and is recommended for all personnel working with VEE (and the related Everglades, Mucambo, Tonate, and Cabassou viruses) or infected animals or entering rooms where these agents or infected animals are present. Work with Hantaan (Korean hemorrhagic fever) virus in rats, voles, and other laboratory rodents should be conducted with special caution (Biosafety Level 4). An inactivated, investigational new Rift Valley fever vaccine (IND) is available from the U.S. Army Medical Research Institute for Infectious Diseases and recommended for all laboratory and animal care personnel working with the agent or infected animals and for all personnel entering laboratories or animal rooms when the agent is in use.

Arboviruses Assigned to Biosafety Level 3

Aino	Israel turkey meningoencephalitis
Akabane	Issyk-kul
Araguari	Itaituba
Batama	Japanese encephalitis
Batken	Kairi
Bhanja	Khasan
Bimbo	Korean hemorrhagic fever (Hantaan)
Bluetongue (exotic)[a]	Koutango
Bobaya	Kyzlagach
Bobia	Louping ill[a]
Buenaventura	Lymphocytic choriomeningitis
Cabassou[c]	Mayaro
Chikungunya[c]	Middelburg
Chim	Mosqueiro
Cocal	Mucambo[c]
Dhori	Murray Valley encephalitis
Dugbe	Nariva
Everglades[c]	Ndumu
Garba	
Germiston[c]	
Getah	
Gordil	
Guaratuba	
Ibaraki	
Ihangapi	
Inini	

Negishi
New Minto
Nodamura
Northway
Oropouche[c]
Orungo
Ouango
Oubangui
Paramushir
Piry
Ponteves
Powassan
Razdan
Rift Valley fever[a,b,c]
Rochambeau
Rocio[c]
Sagiyama
Sakpa
Salanga
Santa Rosa
Saumarez Reef
Semliki Forest
Sepik
Serra do Navio
Slovakia
Spondweni
St. Louis encephalitis
Tamdy
Telok Forest
Thogoto
Tlacotalpan
Tonate[c]
Venezuelan equine encephalomyelitis[c]
VSV-Alagoas
Wesselsbron[a,c]
West Nile
Yellow fever[c]
Zinga[a,b,c]

[a] The importation, possession, or use of this agent is restricted by USDA regulation or administrative policy. See Appendix A.5.
[b] Zinga virus is now recognized as being identical to Rift Valley fever virus.
[c] SALS recommends that work with this agent should be conducted only in Biosafety Level 3 facilities which provide for HEPA filtration of all exhaust air prior to discharge from the laboratory. All persons working with agents for which a vaccine is available should be immunized.

Arboviruses, Arenaviruses, or Filoviruses Assigned to Biosafety Level 4

SALS has recommended that work with the arboviruses, arenaviruses, or filoviruses (63) included in the listing that follows should be conducted at the equivalent of Biosafety Level 4 practices, safety equipment, and facilities. These recommendations are based on documented cases of severe and frequently fatal naturally occurring human infections and aerosol-transmitted laboratory infections. Additionally, SALS recommended that certain agents with a close or identical antigenic relationship to the Biosafety Level 4 agents (e.g., Absettarov and Kumlinge viruses) also be handled at this level until sufficient laboratory experience is obtained to retain these agents at this level or to work with them at a lower level.

Laboratory or laboratory animal-associated infections have been reported with the following agents (37, 53, 58, 67, 90, 112, 12).

Virus	Cases (SALS)
Kyasanur Forest disease	133
Hypr	37 (2 deaths)
Junin	21 (1 death)
Marburg	25 (5 deaths)
Russian spring-summer encephalitis	8
Congo-Crimean hemorrhagic fever	8 (1 death)
Omsk hemmorrhagic fever	5
Lassa	2 (1 death)
Machupo	1 (1 death)
Ebola	1

Rodents are natural reservoirs of Lassa fever virus (*Mastomys natalensis*), Junin and Machupo viruses (*Calomys* spp.), and perhaps other viruses assigned to Biosafety Level 4. Nonhuman primates were associated with the initial outbreaks of Kyasanur Forest disease (*Presbytis* spp.) and Marburg disease (*Cercopithecus* spp.), and arthropods are the natural vectors of the tick-borne encephalitis complex agents. Work with or exposure to rodents, nonhuman primates, or vectors naturally or experimentally infected with these agents represents a potential source of human infection.

Laboratory hazards. The infectious agents may be present in blood, urine, respiratory and throat secretions, semen, and tissues from human or animal hosts and in arthropods, rodents, and nonhuman primates. Respiratory exposure to infectious aerosols, mucous membrane exposure to infectious droplets, and accidental parenteral inoculation are the primary hazards to laboratory or animal care personnel (67, 124).

Recommended precautions. Biosafety Level 4 practices, containment equipment, and facilities are recommended for all activities utilizing known or potentially infectious materials of human, animal, or arthropod origin. Clinical specimens from persons suspected of being infected with one of the agents listed in this summary should be submitted to a laboratory with a Biosafety Level 4 maximum containment facility (22).

Arboviruses, Arenaviruses, and Filoviruses
Assigned to Biosafety Level 4

Congo-Crimean hemorrhagic fever	Marburg
Tick-borne encephalitis virus complex (Absettarov, Hanzalova, Hypr, Kumlinge, Kyasanur Forest disease, Omsk hemorrhagic fever, and Russian spring-summer encephalitis)	Ebola Junin Lassa Machupo

APPENDIX A.1. BIOLOGICAL SAFETY CABINETS

Biological safety cabinets are among the most effective, as well as the most commonly used, primary containment devices in laboratories working with infectious agents. Each of the three types—Class I, II, and III—has performance characteristics which are described below. In addition to the design, construction, and performance standards for vertical laminar flow biological safety cabinets (Class III), the National Sanitation Foundation has also developed a list of such products which meet the reference standard. Utilization of this standard (80) and list should be the first step in selection and procurement of a biological safety cabinet.

Class I and II biological safety cabinets, when used in conjunction with good microbiological techniques, provide an effective partial containment system for safe manipulation of moderate and high-risk microorganisms (i.e., Biosafety Level 2 and 3 agents). Both Class I and II biological safety cabinets have comparable inward face velocities (75 linear ft/min) and provide comparable levels of containment in protecting the laboratory worker and the immediate laboratory environment from infectious aerosols generated within the cabinet.

It is imperative that Class I and II biological safety cabinets are tested and certified in situ at the time of installation within the laboratory, at any time the cabinet is moved, and at least annually thereafter. Certification at locations other than the final site may attest to the performance capability of the individual cabinet or model but does not supersede the critical certification prior to use in the laboratory.

As with any other piece of laboratory equipment, personnel must be trained in the proper use of the biological safety cabinets. The slide-sound training film developed by NIH (Effective Use of the Laminar Flow Biological Safety Cabinet) provides a thorough training and orientation guide. Of particular note are those activities which may disrupt the inward directional airflow through the work opening of Class I and II cabinets. Repeated insertion and withdrawal of the workers' arms in and from the work chamber, opening and closing doors to the laboratory or isolation cubicle, improper placement or operation of materials or equipment within the work chamber, or briskly walking past the biological safety cabinet while it is in use are demonstrated causes of the escape of aerosolized particles from within the cabinet. Strict adherence to recommended practices for the use of biological safety cabinets is as important in attaining the maximum containment capability of the equipment as is the mechanical performance of the equipment itself.

Horizontal laminar flow "clean benches" are present in a number of clinical, pharmacy, and laboratory facilities. These "clean benches" provide a high-quality environment within the work chamber for manipulation of nonhazardous materials. *Caution*: Since the operator sits in the immediate downstream exhaust from the "clean bench," this equipment must never be used for the handling of toxic, infectious, or sensitizing materials.

Class I. The Class I biological safety cabinet (Figure A.1) is an open-fronted, negative-pressure, ventilated cabinet with a minimum inward face velocity at the work opening of at least 75 ft/min. The exhaust air from the cabinet may be used in three operational modes: with a full-width open front, with an installed front closure panel not equipped with gloves, and with an installed front closure panel equipped with arm-length rubber gloves.

Class II. The Class II vertical laminar-flow biological cabinet (Figure A.2) is an open-fronted, ventilated cabinet with an average inward face velocity at the work opening of at least 75 ft/min. This cabinet provides a HEPA-filtered, recirculated mass airflow within the work space. The exhaust air from the cabinet is also filtered by HEPA filters. Design, construction, and performance standards for Class II cabinets have been developed by and are available from the National Sanitation Foundation, Ann Arbor, Michigan (80).

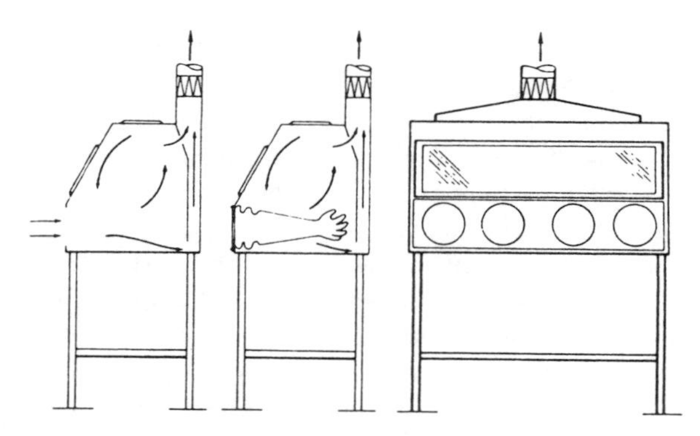

FIGURE A.1 Class I biological safety cabinet

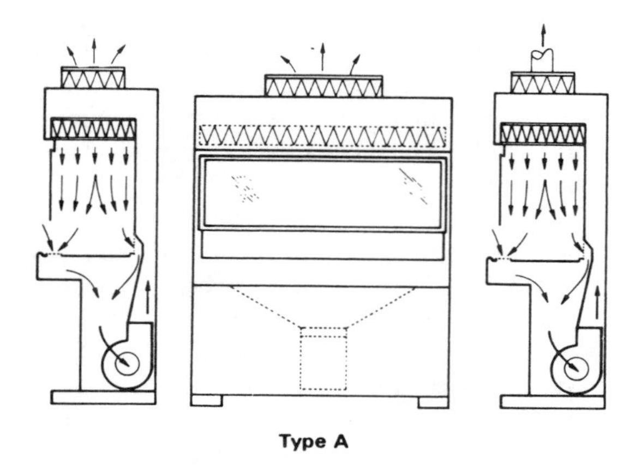

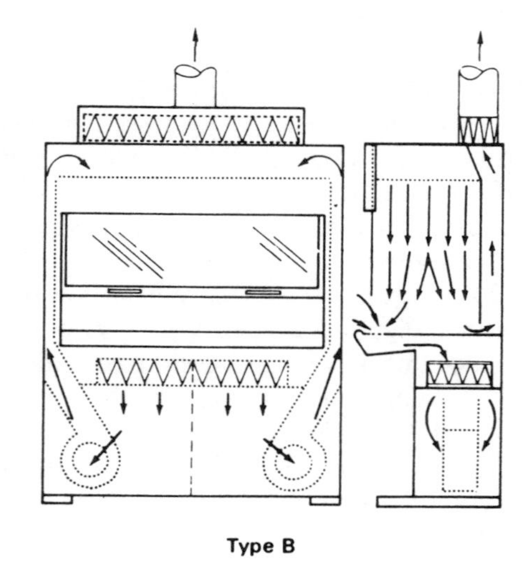

FIGURE A.2 Class II biological safety cabinets

Class III. The Class III cabinet (Figure A.3) is a totally enclosed ventilated cabinet of gas-tight construction. Operations within the Class III cabinet are conducted through attached rubber gloves. When in use, the Class III cabinet is maintained under negative air pressure of at least 0.5 in. water gauge. Supply air is drawn into the cabinet through HEPA filters. The cabinet exhaust air is filtered by two HEPA filters, installed in series, before discharge outside of the facility. The exhaust fan for the Class III cabinet is generally separate from the exhaust fans of the facility's ventilation system.

Use of cabinets. Personnel protection provided by Class I and Class II cabinets is dependent on the inward airflow. Since the face velocities are similar, they generally provide an equivalent level of personnel protection. The use of these cabinets alone, however, is not appropriate for containment of highest-risk infectious agents because aerosols may accidentally escape through the open front.

The use of a Class II cabinet in the microbiological laboratory offers the additional capability and advantage of protecting materials contained within it from extraneous airborne contaminants. This capability is provided by the HEPA-filtered, recirculated mass airflow within the work space.

The Class III cabinet provides the highest level of personnel and product protection. This protection is provided by the physical isolation of the space in which the infectious agent is maintained. When these cabinets are required, all procedures involving infec-

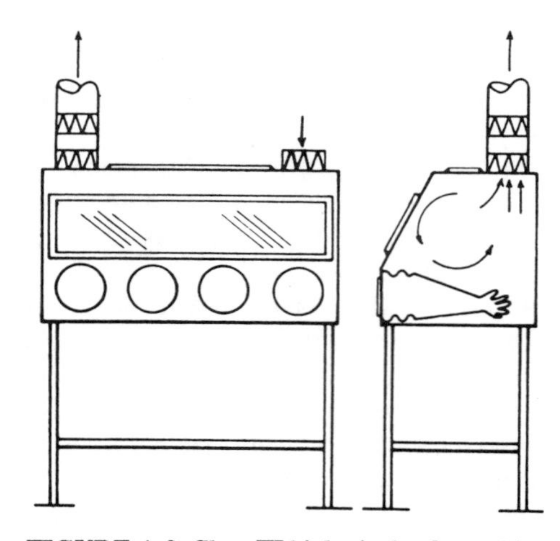

FIGURE A.3 Class III biological safety cabinet

tious agents are contained within them. Several Class III cabinets are therefore typically set up as an interconnected system. All equipment required by the laboratory activity, such as incubators, refrigerators, and centrifuges, must be an integral part of the cabinet system. Double-doored autoclaves and chemical dunk tanks are also attached to the cabinet system to allow supplies and equipment to be safely introduced and removed.

Personnel protection equivalent to that provided by Class III cabinets can also be obtained with a personnel suit area and Class I or Class II cabinets. This area is one in which the laboratory worker is protected from a potentially contaminated environment by a one-piece positive-pressure suit ventilated by a life-support system. This area is entered through an airlock fitted with airtight doors. A chemical shower is provided to decontaminate the surfaces of the suit as the worker leaves the area. The exhaust air from the suit area is filtered by two HEPA units installed in series.

APPENDIX A.2. IMMUNOPROPHYLAXIS

An additional level of protection for at-risk personnel may be achieved with appropriate prophylactic vaccinations. A written organizational policy which defines at-risk personnel, which specifies risks as well as benefits of specific vaccines, and which distinguishes between required and recommended vaccines is essential. In developing such an organizational policy, these recommendations and requirements should be specifically targeted at infectious diseases known or likely to be encountered in a particular facility.

Vaccines for which the benefits (levels of antibody considered to be protective) clearly exceed the risks (local or systemic reactions) should be required for all clearly identified at-risk personnel. Examples of such preparations include vaccines against yellow fever, rabies, and poliomyelitis. Recommendations for giving less efficacious vaccines, those associated with high rates of local or systemic reactions, or those that produce increasingly severe reactions with repeated use should be carefully considered. Products with these characteristics (e.g., cholera, tularemia, and typhoid vaccines) may be recommended but should not ordinarily be required for employment. A complete record of vaccines received on the basis of occupational requirements or recommendations should be maintained in the employee's permanent medical file.

Recommendations for the use of vaccines, adapted from those of the Public Health Service Advisory Committee on Immunization Practices, are included in the agent summary statements in Section V.

APPENDIX A.3. SURVEILLANCE OF PERSONNEL FOR LABORATORY-ASSOCIATED RICKETTSIAL INFECTIONS

Under natural circumstances, the severity of disease caused by rickettsial agents varies considerably. In the laboratory, very large inocula which might produce unusual and perhaps very serious responses are possible. Surveillance of personnel for laboratory-associated infections with rickettsial agents can dramatically reduce the risk of serious consequences of disease.

Recent experience indicates that infections treated adequately with specific anti-rickettsial chemotherapy on the first day of disease do not generally present serious problems. Delay in instituting appropriate chemotherapy, however, may result in debilitating or severe acute disease ranging from increased periods of convalescence in typhus and scrub typhus to death in *R. rickettsii* infections. The key to reducing the severity of disease from laboratory-associated infections is a reliable surveillance system which includes (1) round-the-clock availability of an experienced medical officer, (2) indoctrination of all personnel into the potential hazards of working with rickettsial agents and advantages of early therapy, (3) a reporting system for all recognized overt exposures and accidents, (4) the reporting of all febrile illnesses, especially those associated with headache, malaise, prostration, when no other certain cause exists, and (5) a nonpunitive atmosphere that encourages reporting of any febrile illness.

Rickettsial agents can be handled in the laboratory with minimal real danger to life when an adequate surveillance system complements a staff who are knowledgeable about the hazards of rickettsial infections and who put to use the safeguards recommended in the agent summary statements.

APPENDIX A.4. IMPORTATION AND INTERSTATE SHIPMENT OF HUMAN PATHOGENS AND RELATED MATERIALS

The importation or subsequent receipt of etiologic agents and vectors of human disease is subject to the Public Health Service Foreign Quarantine Regulations (42 CFR, Section 71.56). Permits authorizing the importation or receipt of regulated materials and specifying conditions under which the agent or vector is shipped, handled, and used are issued by the Centers for Disease Control.

The interstate shipment of indigenous etiologic agents, diagnostic specimens, and biological products is subject to applicable packaging, labeling, and shipping requirements of the Interstate Shipment of Etiologic Agents (42 CFR, Part 72). Packaging and labeling requirements for interstate shipment of etiologic agents are summarized and illustrated in Figure A.4.

Additional information on the importation and interstate shipment of etiologic agents of human disease and other related materials may be obtained by writing to:

Centers for Disease Control
Attention: Office of Biosafety
1600 Clifton Road, N.E.
Atlanta, GA 30333
Telephone: (404) 329-3883
FTS: 236-3883

APPENDIX A.5. RESTRICTED ANIMAL PATHOGENS

Nonindigenous pathogens of domestic livestock and poultry may require special laboratory design, operation, and containment features not generally addressed in this publication. The importation, possession, or use of the following agents is prohibited or restricted by law or by U.S. Department of Agriculture regulations or administrative policies:

African horse sickness virus
African swine fever virus
Besnoitia besnoiti
Borna disease virus
Bovine ephemeral fever
Bovine infectious petechial fever agent
Camelpox virus
Foot and mouth disease virus
Fowl plague virus
Histoplasma (Zymonema) farciminosum
Hog cholera virus
Louping ill virus
Lumpy skin disease virus
Mycoplasma agalactiae
Mycoplasma mycoides
Nairobi sheep disease virus (Ganjam virus)
Newcastle disease virus (velogenic strains)
Pseudomonas mallei
Rickettsia ruminantium
Rift Valley fever virus
Rinderpest virus
Swine vesicular disease virus
Teschen disease virus
Theileria annulata
Theileria bovis
Theileria hirci
Theileria lawrencei
Trypanosoma evansi
Trypanosoma vivax
Vesicular exanthema virus
Wesselsbron disease virus

The importation, possession, use, or interstate shipment of animal pathogens other than those listed above may also be subject to regulations of the U.S. Department of Agriculture.

Additional information may be obtained by writing to:

Chief Staff Veterinarian
Organisms and Vectors
Veterinary Services
Animal and Plant Health Inspection Service
U.S. Department of Agriculture
Hyattsville, MD 20782
Telephone: (301) 436-8017
FTS 436-8017

APPENDIX A.6. RESOURCES FOR INFORMATION

Resources for information, consultation, and advice on biohazard control, decontamination procedures, and other aspects of laboratory safety management include:

Centers for Disease Control
Attention: Office of Biosafety
Atlanta, GA 30333
Telephone: (404) 329-3883
FTS 236-3883

National Institutes of Health
Attention: Division of Safety
Bethesda, MD 20892
Telephone: (301) 496-1357
FTS 496-1357

National Animal Disease Center
U.S. Department of Agriculture
Ames, IA 50010
Telephone: (515) 862-8258
FTS 862-8258

LITERATURE CITED

1. Allen, B. W. 1981. Survival and tubercle bacilli in heat-fixed sputum smears. J. Clin. Pathol. 34:719-722.

2. American Thoracic Society Policy Statement. 1974. Quality of laboratory services for mycobacterial disease. Am. Rev. Respir. Dis. 110:376-377.

3. Anonymous. 1980. Tuberculosis infection associated with tissue processing. Calif. Morbid. 30.

4. Anonymous. 1936. Bacteriologist dies of meningitis. J. Am. Med. Assoc. 106:129.

5. Beeman, E. A. 1950. Q fever—an epidemiological note. 1950. Public Health Rep. 65:88-92.

6. Bernard, K. W., G. L. Parham, W. G. Winkler, and C. G. Helmick. 1982. Q fever control measures: recommendations for research of facilities using sheep. Infect. Control 3(6).

7. Blaser, M. J., F. W. Hickman, J. J. Farmer III, D. J. Brenner, A. Balows, and R. A. Feldman. 1980. *Salmonella typhi*: The laboratory as a reservoir of infection. J. Infect. Dis. 142:934-938.

8. Bowen, G. S., C. H. Calisher, W. G. Winkler, A. L. Kraus, E. H. Fowler, R. H. Garman, D. W. Fraser, and A. R. Hinman. 1975. Laboratory studies of a lymphocytic choriomeningitis virus outbreak in man and laboratory animals. Am. J. Epidemiol. 102:233-240.

9. Bozeman, F. M., S. A. Masiello, M. S. Williams, and B. L. Elisberg. 1975. Epidemic typhus rickettsiae isolated from flying squirrels. Nature (London) 255:545-547.

10. Burke, D. S. 1977. Immunization against tularemia: analysis of the effectiveness of live *Francisella tularensis* vaccine in prevention of laboratory-acquired tularemia. J. Infect. Dis. 135:55-60.

11. Burmeister, R. W., W. D. Tigertt, and E. L. Overholt. 1962. Laboratory-acquired pneumonic plague. Ann. Intern. Med. 56:789-800.

12. Bush, J. D. 1943. Coccidioidomycosis. J. Med. Assoc. Alabama 13:159-166.

13. Carougeau, M. 1909. Premier cas Africain de sporotrichose de deBeurmann: transmission de la sporotrichose du mulet a l'homme. Bull. Mem. Soc. Med. Hop. (Paris) 28:507-510.

14. Cavenaugh, D. C., B. L. Elisberg, C. H. Llewellyn, J. D. Marshall, Jr., J. H. Rust, J. E. Williams, and K. F. Meyer. 1974. Plague immunization. IV. Indirect evidence for the efficacy of plague vaccine. J. Infect. Dis. 129(Supplement):S37-S40.

15. Center for Disease Control, Office of Biosafety. 1974. Classification of etiologic agents on the basis of hazard, 4th ed. U.S. Department of Health, Education and Welfare, Public Health Service.

16. Center for Disease Control. 1976. Unpublished data. Center for Infectious Diseases. U.S. Department of Health, Education and Welfare, Public Health Service.

17. Center for Disease Control. 1977. Rabies in a laboratory worker, New York. Morbid. Mortal. Weekly Rep. 26:183-184.

18. Center for Disease Control. 1978. Laboratory-acquired endemic typhus. Morbid. Mortal. Weekly Rep. 27:215-216.

19. Center for Disease Control. 1978. Meningococcal polysaccharide vaccines. Recommendations of the Immunization Practices Advisory Committee (ACIP). Morbid. Mortal. Weekly Rep. 27:327-328.

20. Centers for Disease Control. 1979. Q fever at a university research center—California. Morbid. Mortal. Weekly Rep. 28:333-334.

21. Centers for Disease Control. 1980. Chagas' disease, Kalamazoo, Michigan. Morbid. Mortal. Weekly Rep. 29:147-148.

22. Centers for Disease Control. 1983. Viral hemorrhagic fever: initial management of suspected and confirmed cases. Morbid. Mortal. Weekly Rep. 32(Suppl.):25S-39S.

23. Centers for Disease Control. 1980. Smallpox vaccines. Recommendation of the Immunization Practices Advisory Committee (ACIP). Morbid. Mortal. Weekly Rep. 29:417-420.

24. Centers for Disease Control. 1981. Recommendations of the Immunization Practices Advisory Committee (ACIP). Diphtheria, tetanus, and pertussis. Morbid. Mortal. Weekly Rep. 30:392-396.

25. Centers for Disease Control. 1982. Recommendations of the Immunization Practices Advisory Committee (ACIP). Poliomyelitis prevention. Morbid. Mortal. Weekly Rep. 31:22-26, 31-34.

26. Centers for Disease Control. 1982. Plague vaccine. Selected recommendations of the Public Health Service Advisory Committee on Immunization Practices (ACIP). Morbid. Mortal. Weekly Rep. 31:301-304.

27. Centers for Disease Control. 1982. Recommendations of the Immunizations Practices Advisory Committee (ACIP). Inactivated hepatitis B virus vaccine. Morbid. Mortal. Weekly Rep. 31:317-328.

28. Conant, N. F. 1955. Development of a method for immunizing man against coccidioidomycosis, Third Quarterly Progress Report. Contract DA-18-064-CML-2563, Duke University, Durham, NC. Available from Defense Documents Center, AD 121-600.

29. Davidson, W. L., and K. Hummeler. 1960. B-virus infection in man. Ann. N.Y. Acad. Sci. 85:970-979.

30. Denton, J. F., A. F. DiSalvo, and M. L. Hirsch. 1967. Laboratory-acquired North American blastomycosis. J. Am. Med. Assoc. 199:935-936.

31. Dickson, E. C. 1937. Coccidioides infection: part I. Arch. Intern. Med. 59: 1029-1044.

32. Dickson, E. C. 1937. "Valley fever" of the San Joaquin Valley and fungus coccidioides. Calif. Western Med. 47:151-155.

33. Dickson, E. C., and M. A. Gifford. 1938. Coccidioides infection (coccidioidomycosis). II. The primary type of infection. Arch. Intern. Med. 62:853-871.

34. Diena, B. B., R. Wallace, F. E. Ashton, W. Johnson, and B. Patenaude. 1976. Gonococcal conjunctivitis: accidental infection. Can. Med. Assoc. J. 115:609.

35. Donham, K. J., and J. R. Leninger. 1977. Spontaneous leprosy-like disease in a chimpanzee. J. Infect. Dis. 136:132-136.

36. Dowdle, W. R., and M. A. W. Hattwick. 1977. Swine influenza virus infections in humans. J. Infect. Dis. 136(Suppl.):S386-S389.

37. Edmond, R. T. D., B. Evans, E. T. W. Bowen, and G. Lloyd. 1977. A case of Ebola virus infection. Br. Med. J. 2:541-544.

38. Ellingson, H. V., P. J. Kadull, H. L. Bookwalter, and C. Howe. 1946. Cutaneous anthrax: report of twenty-five cases. J. Am. Med. Assoc. 131:1105-1108.

39. Evans, N. 1903. A clinical report of a case of blastomycosis of the skin from accidental inoculation. J. Am. Med. Assoc. 40:1172-1175.

40. Eyles, D. E., G. R. Coatney, and M. E. Getz. 1960. Vivax-type malaria parasite of macaques transmissible to man. Science 131:1812-1813.

41. Fava, A. 1909. Un cas de sporotrichose conjonctivale et palpebrale primitives. Ann. Ocul. (Paris) 141:338-343.

42. Federal Register. 1976. Recombinant DNA research guidelines. 41:27902-27943.

43. Federal Register. 1982. Guidelines for research involving recombinant DNA molecules. 47:38048-38068.

44. Fielitz, H. 1910. Ueber eine Laboratoriumsinfektion mit dem Sporotrichum de Beurmanni. Zentralbl. Bakteriol. Parasitenkd. Abt. 1 Orig. 55:361-370.

45. Furcolow, M. L. 1961. Airborne histoplasmosis. Bacteriol. Rev. 25:301-309.

46. Gajdusek, D. C., C. J. Gibbs, D. M. Asher, P. Brown, A. Diwan, P. Hoffman, G. Nemo, R. Rohwer, and L. White. 1977. Precautions in the medical care and in handling materials from patients with transmissible virus dementia (Creutzfeldt-Jakob Disease). N. Engl. J. Med. 297:1253-1258.

47. Good, R. C., and E. E. Snider, Jr. 1982. Isolation of nontuberculosis mycobacteria in the U.S. J. Infect. Dis. 146(6).

48. Green, R. N., and P. G. Tuffnell. 1968. Laboratory-acquired melioidosis. Am. J. Med. 44:599-605.

49. Gutteridge, W. E., B. Cover, and A. J. D. Cooke. 1974. Safety precautions for working with *Trypanosoma cruzi*. Trans. R. Soc. Trop. Med. Hyg. 68:161.

50. Halde, C. 1964. Percutaneous *Cryptococcus neoformans* inoculation without infection. Arch. Dermatol. 89:545.

51. Hanel, E., Jr., and R. H. Kruse. 1967. Laboratory-acquired mycoses. Miscellaneous Publication 28. Department of the Army.

52. Hanson, R. P., et al. 1950. Human infections with the virus of vesicular stomatitis. J. Lab. Clin. Med. 36:754-758.

53. Hanson, R. P., S. E. Sulkin, E. L. Buescher, W. McD. Hammond, R. W. McKinney, and T. E. Work. 1967. Arbovirus infections of laboratory workers. Science 158:1283-1286.

54. Harrell, E. R. 1964. The known and the unknown of the occupational mycoses, p. 176-178. *In* Occupational diseases acquired from animals. Continued Education Series no. 124. University of Michigan School of Public Health, Ann Arbor.

55. Harrington, J. M., and H. S. Shannon. 1976. Incidence of tuberculosis, hepatitis, brucellosis and shigellosis in British medical laboratory workers. Br. Med. J. 1:759-762.

56. Hartley, E. G. 1966. "B" virus disease in monkeys and man. Br. Vet. J. 122:46-50.

57. Hattwick, M. A. W., R. J. O'Brien, and B. F. Hanson. 1976. Rocky Mountain spotted fever: epidemiology of an increasing problem. Ann. Intern. Med. 84:732-739.

58. Hennessen, W. 1971. Epidemiology of "Marburg Virus" disease, p. 161-165. *In* G. A. Martini and R. Siegert (ed)., Marburg virus disease. Springer-Verlag, New York.

59. Huddleson, I. F., and M. Munger. 1940. A study of an epidemic of brucellosis due to *Brucella melitensis*. Am. J. Public Health 30:944-954.

60. Jeanselme, E., and P. Chevallier. 1910. Chancres sporotrichosiques des doigts produits par la morsure d'un rat inocule de sporotrichose. Bull. Mem. Soc. Med. Hop. (Paris) 30:176-178.

61. Jeanselme, E., and P. Chevallier. 1911. Transmission de la sporotrichose à l'homme par les morsures d'un rat blanc inocule avec une nouvelle variété de *Sporotrichum*: lymphangite gommeuse ascendante. Bull. Mem. Soc. Med. Hop. (Paris) 31:287-301.

62. Kaufmann, A. F., and D. C. Anderson. 1978. Tuberculosis control in nonhuman primates, p. 227-234. *In* R. J. Montali (ed.), Mycobacterial infections of zoo animals. Smithsonian Institution Press, Washington, D.C.

63. Kiley, M. P., E. T. W. Bowel, G. A. Eddy, M. Isaacson, K. M. Johnson, J. B. McCormick, F. A. Murphy, S. R. Pattyn, D. Peters, W. Prozesky, R. Regnery, D. I. H. Simpson, W. Slenczka, P. Sureau, G. van der Groen, P. A. Webb, and H. Sulff. 1982. Filoviridae: taxonomic home for Marburg and Ebola viruses? Intervirology 18:24-32.

64. Klutsch, K., N. Hummer, U. Braun, A. Heidland. 1965. Zur Klinik der Coccidioidomykose. Dtsch. Med. Wochenschr. 90:1498-1501.

65. Kubica, G. P., W. Gross, J. E. Hawkins, H. M. Sommers, A. L. Vestal, and L. G. Wayne. 1975. Laboratory services for mycobacterial diseases. Am. Rev. Respir. Dis. 112:773-787.5.

66. Larsh, H. W., and J. Schwartz. 1977. Accidental inoculation— blastomycosis. Cutis 19:334-336.

67. Leifer, E., D. J. Gocke, and H. Bourne. 1970. Lassa fever, a new virus disease of man from West Africa. II. Report of a laboratory-acquired infection treated with plasma from a person recently recovered from the disease. Am. J. Trop. Med. Hyg. 19:677-679.

68. Looney, J. M., and T. Stein. 1950. Coccidioidomycosis. N. Engl. J. Med. 242:77-82.

69. Marchoux, P. E. 1934. Un cas d'inoculation accidentelle du bacille de Hanson en pays non lepreux. Int. J. Leprosy 2:1-7.

70. Martini, G. A., and H. A. Schmidt. 1968. Spermatogenic transmission of Marburg virus. Klin. Wochenschr. 46:398-400.

71. McAleer, R. 1980. An epizootic in laboratory guinea pigs due to *Trichophyton mentagrophytes*. Austral. Vet. J. 56:234-236.

72. McDade, J. E., and C. C. Shepard. 1979. Virulent to avirulent conversion of Legionnaire's disease bacterium (*Legionella pneumophila*). Its effect on isolation techniques. J. Infect. Dis. 139:707-711.

73. Melnick, J. L., H. A. Wenner, and C. A. Phillips. 1979. Enteroviruses, p. 471-534. *In* E. H. Lennette and N. J. Schmidt (ed.), Diagnostic procedures for viral, rickettsial and chlamydial infections, 5th ed. American Public Health Association, Washington, D.C.

74. Meyer, K. F. 1915. The relationship of animal to human sporotrichosis: studies on American sporotrichosis III. J. Am. Med. Assoc. 65:579-585.

75. Meyer, K. F., and B. Eddie. 1941. Laboratory infections due to *Brucella*. J. Infect. Dis. 68:24-32.

76. Meyers, W. M., G. P. Walsh, H. L. Brown, Y. Fukunishi, C. H. Binford, P. J. Gerone, and R. H. Wolf. 1980. Naturally acquired leprosy in a mangabey monkey (*Cercocebus* sp.). Int. J. Leprosy 48:495-496.

77. Morisset, R., and W. W. Spink. 1969. Epidemic canine brucellosis due to a new species, *Brucella canis*. Lancet ii:1000-1002.

78. Murray, J. F., and D. H. Howard. 1964. Laboratory-acquired histoplasmosis. Am. Rev. Respir. Dis. 89:631-640.

79. Nabarro, J. D. N. 1948. Primary pulmonary coccidioidomycosis: case of laboratory infection in England. Lancet i:982-984.

80. National Sanitation Foundation. 1983. Standard 49. Class II (laminar flow) biohazard cabinetry. National Sanitation Foundation, Ann Arbor, Mich.

81. Norden, A. 1951. Sporotrichosis: clinical and laboratory features and a serologic study in experimental animals and humans. Acta Pathol. Microbiol. Scand. Suppl. 89:3-119.

82. Office of Research Safety, National Cancer Institute, and the Special Committee of Safety and Health Experts. 1978. Laboratory safety monograph: a supplement to the NIH Guidelines for Recombinant DNA Research. National Institutes of Health, Bethesda, Maryland.

83. Oliphant, J. W., and R. R. Parker. 1948. Q fever: three cases of laboratory infection. Public Health Rep. 63:1364-1370.

84. Oliphant, J. W., D. A. Gordon, A. Meis, and R. R. Parker. 1949. Q fever in laundry workers, presumably transmitted from contaminated clothing. Am. J. Hyg. 49:76-82.

85. Oliver, J., and T. R. Reusser. 1942. Rapid method for the concentration of tubercle bacilli. Am. Rev. Tuberc. 45:450-452.

86. Oster, C. N., et al. 1977. Laboratory-acquired Rocky Mountain spotted fever. The hazard of aerosol transmission. N. Engl. J. Med. 297:859-862.

87. Parritt, R. J., and R. E. Olsen. 1947. Two simultaneous cases of leprosy developing in tattoos. Am. J. Pathol. 23:805-817.

88. Patterson, W. C., L. O. Mott, and E. W. Jenney. 1958. A study of vesicular stomatitis in man. J. Am. Vet. Med. Assoc. 133:57-62.

89. Perkins, F. T., and E. G. Hartley. 1966. Precautions against B virus infection. Br. Med. J. 1:899-901.

90. Pike, R. M. 1976. Laboratory-associated infections: summary and analysis of 3,921 cases. Health Lab. Sci. 13:105-114.

91. Pike, R. M. 1978. Past and present hazards of working with infectious agents. Arch. Pathol. Lab. Med. 102:333-336.

92. Pike, R. M. 1979. Laboratory-associated infections: incidence, fatalities, causes and prevention. Annu. Rev. Microbiol. 33:41-66.

93. Pike, R. M., S. E. Sulkin, and M. L. Schulze. 1965. Continuing importance of laboratory-acquired infections. Am. J. Public Health 55:190-199.

94. Prescott, J. F., and M. A. Karmali. 1978. Attempts to transmit *Campylobacter enteritis* to dogs and cats. Can. Med. Assoc. J. 119:1001-1002.

95. Reid, D. D. 1957. Incidence of tuberculosis among workers in medical laboratories. Br. Med. J. 2:10-14.

96. Report of the Committee of Inquiry into the Smallpox Outbreak in London in March and April 1973. 1974. Her Majesty's Stationery Office, London.

97. Richardson, J. H. 1973. Provisional summary of 109 laboratory-associated infections at the Center for Disease Control, 1947-1973. Presented at the 16th Annual Biosafety Conference, Ames, Iowa.

98. Riley, R. L. 1957. Aerial dissemination of pulmonary tuberculosis. Am. Rev. Tuberc. 76:931-941.

99. Riley, R. L. 1961. Airborne pulmonary tuberculosis. Bacteriol. Rev. 25:243-248.

100. Robertson, D. H. H., S. Pickens, J. H. Lawson, and B. Lennex. 1980. An accidental laboratory infection with African trypanosomes of a defined stock. I and II. J. Infect. Dis. 2:105-112, 113-124.

101. Sastaw, S., and H. N. Carlisle. 1966. Aerosol infection of monkeys with *Rickettsia rickettsii*. Bacteriol. Rev. 30:636-645.

102. Schlech, W. F., J. B. Turchik, R. E. Westlake, G. C. Klein, J. D. Band, and R. E. Wever. 1981. Laboratory-acquired infection with *Pseudomonas pseudomallei* (melioidosis). N. Engl. J. Med. 305:1133-1135.

103. Schwarz, J., and G. L. Baum. 1951. Blastomycosis. Am. J. Clin. Pathol. 21:999-1029.

104. Skinhoj, P. 1974. Occupational risks in Danish clinical chemical laboratories. II. Infections. Scand. J. Clin. Lab. Invest. 33:27-29.

105. Smith, C. E. 1950. The hazard of acquiring mycotic infections in the laboratory. Presented at 78th Annual Meeting, American Public Health Association, St. Louis, Missouri.

106. Smith, C. E., D. Pappagianis, H. B. Levine, and M. Saito. 1961. Human coccidioidomycosis. Bacteriol. Rev. 25:310-320.

107. Smith, D. T., and E. R. Harrell, Jr. 1948. Fatal coccidioidomycosis: a case of laboratory infection. Am. Rev. Tuberc. 57:368-374.

108. Smithwick, R. W., and C. B. Stratigos. 1978. Preparation of acid-fast microscopy smears for proficiency testing and quality control. J. Clin. Microbiol. 8:110-111.

109. Spinelli, J. S., et al. 1981. Q fever crisis in San Francisco: controlling a sheep zoonosis in a lab animal facility. Lab. Anim. 10:24-27.

110. Spink, W. W. 1956. The nature of brucellosis, p. 106-108. University of Minnesota Press, Minneapolis.

111. Sterne, M., and L. M. Wertzel. 1950. A new method of large-scale production of high-titer botulinum formol-toxoid types C and D. J. Immunol. 65:175-183.

112. Subcommittee on Arbovirus Laboratory Safety of the American Committee on Arthropod-borne Viruses. 1980. Laboratory safety for arboviruses and certain other viruses of vertebrates. Am. J. Trop. Med. Hyg. 29:1359-1381.

113. Sulkin, S. E., and R. M. Pike. 1949. Viral infections contracted in the laboratory. N. Engl. J. Med. 241:205-213.

114. Sulkin, S. E., and R. M. Pike. 1951. Survey of laboratory-acquired infections. Am. J. Public Health 41:769-781.

115. Sullivan, J. F., J. R. Songer, and I. E. Estrem. 1978. Laboratory-acquired infections at the National Animal Disease Center, 1960-1976. Health Lab. Sci. 15:58-64.

116. Tesh, R. B., and J. D. Schneidau, Jr. 1966. Primary cutaneous histoplasmosis. N. Engl. J. Med. 275:597-599.

117. Thompson, D. W., and W. Kaplan. 1977. Laboratory-acquired sporotrichosis. Sabouraudia 15:167-170.

118. Tomlinson, C. C., and P. Bancroft. 1928. Granuloma coccidioides: report of a case responding favorably to antimony and potassium tartrate. J. Am. Med. Assoc. 91:947-951.

119. Tosh, F. E., J. Balhuizen, J. L. Yates, and C. A. Brasher. 1964. Primary cutaneous histoplasmosis: report of a case. Arch. Intern. Med. 114:118-119.

120. Walsh, G. P., E. E. Storrs, H. P. Burchfield, E. H. Cottrel, M. F. Vidrine, and C. H. Binford. 1975. Leprosy-like disease occurring naturally in armadillos. J. Reticuloendothel. Soc. 18:347-351.

121. Wedum, A. G., and R. H. Kruse. 1969. Assessment of risk of human infection in the microbiology laboratory. Miscellaneous Publication no. 30, Industrial Health and Safety Directorate, Fort Detrick, Frederick, Maryland.

122. Wedum, A. G., W. E. Barkley, and A. Hellman. 1972. Handling of infectious agents. J. Am. Vet. Med. Assoc. 161:1557-1567.

123. Wedum, A. G. 1975. History of microbiological safety. 18th Biological Safety Conference. Lexington, Kentucky.

124. Weissenbacher, M. C., M. E. Grela, M. S. Sabattini, J. I. Maiztegui, C. E. Coto, M. J. Frigerio, P. M. Cossio, A. S. Rabinovich, and J. G. B. Oro. 1978. Inapparent infections with Junin virus among laboratory workers. J. Infect. Dis. 137:309-313.

125. Wilder, W. H., and C. P. McCullough. 1914. Sporotrichosis of the eye. J. Am. Med. Assoc. 62:1156-1160.

126. Willems, W. R., G. Kaluza, C. B. Boschek, and H. Bauer. 1979. Semliki Forest virus: cause of a fatal case of human encephalitis. Science 203:1127-1129.

127. Wilson, J. W., E. P. Cawley, F. D. Weidman, and W. S. Gilmer. 1955. Primary cutaneous North American blastomycosis. Arch. Dermatol. 71:39-45.

128. Wilson, J. W., C. E. Smith, and O. A. Plunkett. 1953. Primary cutaneous coccidioidomycosis; the criteria for diagnosis and a report of a case. Calif. Med. 79:233-239.

129. Winkler, W. G. 1973. Airborne rabies transmission in a laboratory worker. J. Am. Med. Assoc. 226:1219-1221.

130. World Health Organization. 1978. Smallpox surveillance. Weekly Epidemiol. Rec. 53:265-266.

Appendix B

1988 Agent Summary Statement for HIVs, Including HTLV-III, LAV, HIV-1, and HIV-2

CONTENTS

Reprinted from *Morbidity and Mortality Weekly Report,* 1988; 37(no. S-4):1-17. The information and recommendations contained in this appendix were developed and compiled by (1) the Division of Safety, National Institute of Allergy and Infectious Diseases, the National Cancer Institute, and the Clinical Center of the National Institutes of Health; (2) the Food and Drug Administration; and (3) the following units of the Centers for Disease Control: AIDS Program, Hospital Infections Program, and Office of the Director, Center for Infectious Diseases; the Training and Laboratory Program Office; and the Office of Biosafety, Office of the Centers Director. Representatives of the following organizations also collaborated in the effort: American Academy of Microbiology, American Biological Safety Association, American Society for Microbiology, American Society for Clinical Pathology, Association of State and Territorial Public Health Laboratory Directors, College of American Pathologists, Pharmaceutical Manufacturers Association, and Walter Reed Army Institute for Research.

1988 AGENT SUMMARY STATEMENT FOR HIVs, INCLUDING HTLV-III, LAV, HIV-1, AND HIV-2

INTRODUCTION

In 1984, the Centers for Disease Control (CDC) and the National Institutes of Health (NIH), in consultation with experts from academic institutions, industry, and government, published the book *Biosafety in Microbiological and Biomedical Laboratories* ("Guidelines")* (1).

These Guidelines are based on combinations of standard and special practices, equipment, and facilities recommended for use in working with infectious agents in various laboratory settings. The recommendations are advisory; they provide a general code for operating microbiologic and biomedical laboratories.

One section of the Guidelines is devoted to standard and special microbiologic practices, safety equipment, and facilities for biosafety levels (BSL) 1 through 4. Another section contains specific agent summary statements, each consisting of a brief description of laboratory-associated infections, the nature of laboratory hazards, and recommended precautions for working with the causative agent. The authors realized that the discovery of the availability of information about these agents would necessitate updating the agent summary. Such a statement for human immunodeficiency virus (HIV) (called HTLV-III/LAV when the Guidelines were published) was published in *Morbidity and Mortality Weekly Report* (MMWR) in 1986 (2). The HIV agent summary statement printed in this Supplement updates the 1986 statement.

Attached to the updated HIV agent summary statement are the essential elements for BSL 2 and 3 laboratories, reproduced from the Guidelines (1) (see Addendum 1, p. 145). BSL 2 and 3 laboratory descriptions are included because they are recommended for laboratory personnel working with HIV, depending on the concentration or quantity of virus or the type of laboratory procedures used.

The HIV agent summary statement does not specifically address safety measures for collecting and handling clinical specimens. Nonetheless, it has been recommended that blood and body-fluid precautions consistently be used for ALL specimens from ALL patients. This approach, referred to as "universal blood and body-fluid precautions" or "universal precautions," eliminates the need to identify all patients infected with HIV (or other blood-borne pathogens) (3). This subject is also covered in other publications (3-8).

Laboratory directors, supervisors, and others are asked to attach a copy of this revised "1988 Agent Summary Statement for HIVs" to each copy of the Guidelines and to all copies of their laboratory biosafety manual; they should review the recommended precautions with laboratory personnel, provide appropriate training in practices and operation of facilities, and ensure that all personnel demonstrate proficiency BEFORE being allowed to work with HIV. The laboratory director (or the designated laboratory supervisor) is responsible for biosafety in the laboratory and must establish and implement practices, facilities, equipment, training, and work assignments as appropriate (9).

HIV AGENT SUMMARY STATEMENT AGENT: HIVS INCLUDING HTLV-III, LAV, HIV-1, AND HIV-2

In the period 1984-1986, several health-care workers (HCWs) who had no recognized risk behavior for acquired immunodeficiency syndrome (AIDS) were reported to have HIV infection (10-15). Only one of these HCWs was identified as a laboratory worker. These and other reports assessed the risk of work-related HIV infection for all HCWs as being very low (3,6,10-12,14-18).

---

*Available from:

Superintendent of Documents
U.S. Government Printing Office
Washington, DC 20402
Stock #01702300167-1

National Technical Information Service
U.S. Department of Commerce
5282 Port Royal Road
Springfield, VA 22161
Stock #PB84-206879

In 1985, anecdotal reports were received indicating that workers in two different HIV-reagent-production laboratories had been exposed to droplets and splashed liquid from a vessel containing concentrated virus. One of several workers had been cut by glass from a broken carboy that contained HIV-infected cells and medium. None of the persons exposed in these episodes had developed antibody to HIV or had clinical signs of infection 18 and 20 months, respectively, after the reported exposure.

In 1987, CDC received reports that three HCWs had HIV infection; none of the infections were associated with needlesticks or cuts. Two of these HCWs were clinical laboratory workers (11). One was a phlebotomist whose face and mouth were splattered with a patient's blood when the rubber stopper was suddenly expelled from a blood-collection tube. The second was a medical technologist who inadvertently spilled blood on her arms and forearms while using an apheresis apparatus to process blood from an HIV-seropositive patient.

In September 1987, a production-laboratory worker was reported to have HIV infection (18). This person worked with large concentrations of HIV in a BSL 3 facility. HIV was isolated from the worker's blood; the isolate was genetically indistinguishable from the strain of virus being cultivated in the laboratory. No risk factors were identified, and the worker recalled no specific incident that might have led to infection. However, there were instances of leakage of virus-positive culture fluid from equipment and contamination of the work area and centrifuge rotors. The report concluded that the most plausible source of exposure was contact of the worker's gloved hand with virus-culture supernatant, followed by inapparent exposure to skin.

In October 1987, a second person who worked in another HIV production facility was reported to have HIV infection (18). This laboratory was a well-equipped BSL 3 facility, and BSL 3 practices were being followed. This worker reported having sustained a puncture wound to a finger while cleaning equipment used to concentrate HIV.

LABORATORY HAZARDS

Human immunodeficiency virus has been isolated from blood, semen, saliva, tears, urine, cerebrospinal fluid, amniotic fluid, breast milk, cervical secretions, and tissue of infected persons and experimentally infected nonhuman primates. In the laboratory, virus should be presumed to be present in all HIV cultures, in all materials derived from HIV cultures, and in/on all equipment and devices coming into direct contact with any of these materials.

In the laboratory, the skin (especially when scratches, cuts, abrasions, dermatitis, or other lesions are present) and mucous membranes of the eye, nose, mouth, and possibly the respiratory tract should be considered as potential pathways for entry of virus. Needles, sharp instruments, broken glass, and other sharp objects must be carefully handled and properly discarded. Care must be taken to avoid spilling and splashing infected cell-culture liquid and other virus-containing materials.

RECOMMENDED PRECAUTIONS

1. BSL 2 standards and special practices, containment equipment, and facilities, as described in the CDC-NIH publication *Biosafety in Microbiological and Biomedical Laboratories* ("Guidelines"), are recommended for activities involving all clinical specimens, body fluids, and tissues from humans or from infected or inoculated laboratory animals. These are the same standards and practices recommended for handling all clinical specimens. For example, and for emphasis:

    a. Use of syringes, needles, and other sharp instruments should be avoided if possible. Used needles and disposable cutting instruments should be discarded into a puncture-resistant container with a lid. Needles should not be re-sheathed, bent, broken, removed from disposable syringes, or otherwise manipulated by hand.

    b. Protective gloves should be worn by all personnel engaged in activities that may involve direct contact of skin with potentially infectious specimens, cultures, or tissues. Gloves should be carefully removed and changed when they are visibly contaminated. Personnel who have dermatitis or other lesions on the hands and who may have indirect contact with potentially infectious material should also wear protective gloves. Hand washing with soap and water immediately after infectious materials are handled and after work is completed—EVEN WHEN GLOVES HAVE BEEN WORN as described above—should be a routine practice.

c. Generation of aerosols, droplets, splashes, and spills should be avoided. A biological safety cabinet should be used for all procedures that might generate aerosols or droplets and for all infected cell-culture manipulations. The Guidelines (pp. 11-13) contain additional precautions for operating at BSL 2.

2. Activities such as producing research-laboratory-scale amounts of HIV, manipulating concentrated virus preparations, and conducting procedures that may produce aerosols or droplets should be performed in a BSL 2 facility with the additional practices and containment equipment recommended for BSL 3 (19) (Guidelines, pp. 14-17).

3. Activities involving industrial-scale, large-volume production or high concentration and manipulation of concentrated HIV should be conducted in a BSL 3 facility using BSL 3 practices and equipment (19).

4. BSL 2 practices, containment equipment, and facilities for animals are recommended for activities involving nonhuman primates and any animals experimentally infected or inoculated with HIV. Because laboratory animals may bite, throw feces or urine, or expectorate at humans, animal-care personnel, investigators, technical staff, and other persons who enter the animal rooms should wear coats, protective gloves, coveralls or uniforms, and—as appropriate—face shields or surgical masks and eye shields to protect the skin and mucous membranes of the eyes, nose, and mouth.

5. All laboratory glassware, disposable material, and waste material suspected or known to contain HIV should be decontaminated, preferably in an autoclave, before it is washed, discarded, etc. An alternate method of disposing of solid wastes is incineration.

6. Laboratory workers should wear laboratory coats, gowns, or uniforms when working with HIV or with material known or suspected to contain HIV. There is no evidence that laboratory clothing poses a risk for HIV transmission; however, clothing that becomes contaminated with HIV preparations should be decontaminated before being laundered or discarded. Laboratory personnel must remove laboratory clothing before going to nonlaboratory areas.

7. Work surfaces should be decontaminated with an appropriate chemical germicide after procedures are completed, when surfaces are overtly contaminated, and at the end of each work day. Many commercially available chemical disinfectants (5,20-23) can be used for decontaminating laboratory work surfaces, for some laboratory instruments, for spot cleaning of contaminated laboratory clothing, and for spills of infectious materials. Prompt decontamination of spills should be standard practice.

8. Universal precautions are recommended for handling all human blood specimens for hematologic, microbiologic, chemical, and serologic testing; these are the same precautions for preventing transmission of all bloodborne infections, including hepatitis B (17,21,24,25). It is not certain how effective 56°C-60°C heat is in destroying HIV in serum (22,23,26), but heating small volumes of serum for 30 minutes at 56°C before serologic testing reduces residual infectivity to below detectable levels. Such treatment causes some false-positive results in HIV enzyme immunoassays (27-30) and may also affect some biochemical assays performed on serum (27,31,32).

9. Human serum from any source that is used as a control or reagent in a test procedure should be handled at BSL 2 (Guidelines, pp. 11-13). Addendum 2 (p. 152) to this report is a statement issued by CDC on the use of all human control and reagent serum specimens shipped to other laboratories. The Food and Drug Administration requires that manufacturers of human serum reagents use a similarly worded statement.

10. Medical surveillance programs should be in place in all laboratories that test specimens, do research, or produce reagents involving HIV. The nature and scope of a surveillance program will vary according to institutional policy and applicable local, state, and Federal regulations (9).

11. If a laboratory worker has a parenteral or mucous-membrane exposure to blood, body fluid, or viral-culture material, the source material should be identified and, if possible, tested for the presence of virus. If the source material is positive for HIV antibody, virus, or antigen, or is not available for examination, the worker should be counseled regarding the risk of infection and should be evaluated clinically and serologically for evidence of HIV infection. The worker should be advised to report on and to seek medical evaluation of any acute febrile illness that occurs within 12 weeks after the exposure (3). Such an illness—particularly one characterized by fever, rash, or lymphadenopathy—may indicate recent HIV

infection. If seronegative, the worker should be retested 6 weeks after the exposure and periodically thereafter (e.g., at 12 weeks and 6 months after exposure). During this follow-up period—especially during the first 6-12 weeks after exposure, when most infected persons are expected to show serologic evidence of infection—exposed workers should be counseled to follow Public Health Service recommendations for preventing transmission of HIV (3,14,25,33). It is recommended that all institutions establish written policies regarding the management of laboratory exposure to HIV; such policies should deal with confidentiality, counseling, and other related issues.

12. Other primary and opportunistic pathogenic agents may be present in the body fluids and tissues of persons infected with HIV. Laboratory workers should follow accepted biosafety practices to ensure maximum protection against inadvertent laboratory exposure to agents that may also be present in clinical specimens (34-36).

13. Unless otherwise dictated by institutional policy, the laboratory director (or designated laboratory supervisor) is responsible for carrying out the biosafety program in the laboratory. In this regard, the laboratory director or designated supervisor should establish the biosafety level for each component of the work to be done and should ensure that facilities and equipment are adequate and in good working order, that appropriate initial and periodic training is provided to the laboratory staff, and that recommended practices and procedures are strictly followed (9).

14. Attention is directed to a "Joint Advisory Notice" of the Departments of Labor and Health and Human Services (9) that describes the responsibility of employers to provide "safe and healthful working conditions" to protect employees against occupational infection with HIV. The notice defines three exposure categories of generic job-related tasks and describes the protective measures required for employees involved in each exposure category. These measures are: administrative measures, training and education programs for employees, engineering controls, work practices, medical and health-care practices, and record-keeping. The recommendations in this report are consistent with the "Joint Advisory Notice"; managers/directors of all biomedical laboratories are urged to read this notice.

ADDENDUM 1

LABORATORY BIOSAFETY LEVEL CRITERIA

Biosafety Level 2

Biosafety Level 2 is similar to Level 1 and is suitable for work involving agents that represent a moderate hazard for personnel and the environment. It differs in that

a. laboratory personnel have specific training in handling pathogenic agents and are directed by competent scientists,
b. access to the laboratory is limited when work is being conducted, and
c. certain procedures in which infectious aerosols are created are conducted in biological safety cabinets or other physical containment equipment.

The following standard and special practices, safety equipment, and facilities apply to agents assigned to Biosafety Level 2:

A. *Standard microbiological practices*

1. Access to the laboratory is limited or restricted by the laboratory director when work with infectious agents is in progress.

2. Work surfaces are decontaminated at least once a day and after any spill of viable material.

3. All infectious liquid or solid waste is decontaminated before being disposed of.

4. Mechanical pipetting devices are used; mouth pipetting is prohibited.

5. Eating, drinking, smoking, and applying cosmetics are not permitted in the work area. Food must be stored in cabinets or refrigerators designed and used for this purpose only. Food storage cabinets or refrigerators should be located outside the work area.

6. Persons are to wash their hands when they leave the laboratory after handling infectious material or animals.

7. All procedures are performed carefully to minimize the creation of aerosols.

B. *Special practices*

1. Contaminated materials that are to be decontaminated away from the laboratory are placed in a durable, leakproof container that is closed before being removed from the laboratory.

2. The laboratory director limits access to the laboratory. In general, persons who are at increased risk of acquiring infection or for whom infection may be unusually hazardous are not allowed in the laboratory or animal rooms. The director has the final responsibility for assessing each circumstance and determining who may enter or work in the laboratory.

3. The laboratory director establishes policies or procedures whereby only persons who have been advised of the potential hazard and who meet any specific entry requirements (e.g., vaccination) enter the laboratory or animal rooms.

4. When an infectious agent being worked with in the laboratory requires special provisions for entry (e.g., vaccination), a hazard warning sign that incorporates the universal biohazard symbol is posted on the access door to the laboratory work area. The hazard warning sign identifies the infectious agent, lists the name and telephone number of the laboratory director or other responsible person(s), and indicates the special requirement(s) for entering the laboratory.

5. An insect and rodent control program is in effect.

6. Laboratory coats, gowns, smocks, or uniforms are worn while in the laboratory. Before leaving the laboratory for nonlaboratory areas (e.g., cafeteria, library, administrative offices), this protective clothing is removed and left in the laboratory or covered with a clean coat not used in the laboratory.

7. Animals not involved in the work being performed are not permitted in the laboratory.

8. Special care is taken to avoid having skin be contaminated with infectious material; gloves should be worn when handling infected animals and when skin contact with infectious material is unavoidable.

9. All waste from laboratories and animal rooms is appropriately decontaminated before disposal.

10. Hypodermic needles and syringes are used only for parenteral injection and aspiration of fluids from laboratory animals and diaphragm bottles. Only needle-locking syringes or disposable syringe-needle units (i.e., the needle is integral to the syringe) are used for the injection or aspiration of infectious fluid. Extreme caution should be used when handling needles and syringes to avoid autoinoculation and the generation of aerosols during use and disposal. A needle should not be bent, sheared, replaced in the sheath or guard, or removed from the syringe following use. The needle and syringe should be promptly placed in a puncture-resistant container and decontaminated, preferably by autoclaving, before discard or reuse.

11. Spills and accidents that result in overt exposures to infectious material are immediately reported to the laboratory director. Medical evaluation, surveillance, and treatment are provided as appropriate, and written records are maintained.

12. When appropriate, considering the agent(s) handled, baseline serum samples for laboratory and other at-risk personnel are collected and stored. Additional serum specimens may be collected periodically, depending on the agents handled or on the function of the facility.

13. A biosafety manual is prepared or adopted. Personnel are advised of special hazards and are required to read instructions on practices and procedures and to follow them.

C. *Containment equipment*

Biological safety cabinets (Class I or II) or other appropriate personal-protection or physical-containment devices are used when:

1. Procedures with a high potential for creating infectious aerosols are conducted. These may include centrifuging, grinding, blending, vigorous shaking or mixing, sonic disruption, opening containers of infectious materials whose internal pressures may be different from ambient pressures, inoculating animals intranasally, and harvesting infected tissues from animals or eggs.

2. High concentrations or large volumes of infectious agents are used. Some types of materials may be centrifuged in the open laboratory if sealed heads or centrifuge safety cups are used and if the containers are opened only in a biological safety cabinet.

D. *Laboratory facilities*

1. The laboratory is designed so that it can be easily cleaned.

2. Bench tops are impervious to water and resistant to acids, alkalis, organic solvents, and moderate heat.

3. Laboratory furniture is sturdy, and spaces between benches, cabinets, and equipment are accessible for cleaning.

4. Each laboratory contains a sink for handwashing.

5. If the laboratory has windows that open, they are fitted with fly screens.

6. An autoclave for decontaminating infectious laboratory wastes is available.

Biosafety Level 3

Biosafety Level 3 is applicable to clinical, diagnostic, teaching, research, or production facilities in which work is done with indigenous or exotic agents that may cause serious or potentially lethal disease as a result of exposure by inhalation. Laboratory personnel have specific training in handling pathogenic and/or potentially lethal agents and are supervised by competent scientists who are experienced in working with these agents. All procedures involving the manipulation of infectious material are conducted within biological safety cabinets or other physical containment devices or by personnel wearing appropriate personal-protection clothing and devices. The laboratory has special engineering and design features. It is recognized, however, that many existing facilities may not have all the facility safeguards recommended for Biosafety Level 3 (e.g., access zone, sealed penetrations, and directional airflow). In these circumstances, acceptable safety may be achieved for routine or repetitive operations (e.g., diagnostic procedures involving the propagation of an agent for identification, typing, and susceptibility testing) in laboratories in which facility features satisfy Biosafety Level 2 recommendations if the recommended "Standard Microbiological Practices," "Special Practices," and "Containment Equipment" for Biosafety Level 3 are rigorously followed. The decision to implement this modification of Biosafety Level 3 recommendations should be made only by the laboratory director.

The following standard and special practices, safety equipment, and facilities apply to agents assigned to Biosafety Level 3:

A. *Standard microbiological practices*

1. Work surfaces are decontaminated at least once a day and after any spill of viable material.

2. All infectious liquid or solid waste is decontaminated before being disposed of.

3. Mechanical pipetting devices are used; mouth pipetting is prohibited.

4. Eating, drinking, smoking, storing food, and applying cosmetics are not permitted in the work area.

5. Persons wash their hands after handling infectious materials and animals and every time they leave the laboratory.

6. All procedures are performed carefully to minimize the creation of aerosols.

B. *Special practices*

1. Laboratory doors are kept closed when experiments are in progress.

2. Contaminated materials that are to be decontaminated at a site away from the laboratory are placed in a durable, leakproof container that is closed before being removed from the laboratory.

3. The laboratory director controls access to the laboratory and limits access only to persons whose presence is required for program or support purposes. Persons who are at increased risk of acquiring infection or for whom infection may be unusually hazardous are not allowed in the laboratory or animal rooms. The director has the final responsibility for assessing each circumstance and determining who may enter or work in the laboratory.

4. The laboratory director establishes policies and procedures whereby only persons who have been advised of the potential biohazard, who meet any specific entry requirements (e.g., vaccination), and who comply with all entry and exit procedures enter the laboratory or animal rooms.

5. When infectious materials or infected animals are present in the laboratory or containment module, a hazard warning sign (incorporating the universal biohazard symbol) is posted on all laboratory and animal-room access doors. The hazard warn-

ing sign identifies the agent, lists the name and telephone number of the laboratory director or other responsible person(s), and indicates any special requirements for entering the laboratory, such as the need for vaccinations, respirators, or other personal-protection measures.

6. All activities involving infectious materials are conducted in biological safety cabinets or other physical-containment devices within the containment module. No work is conducted in open vessels on the open bench.

7. The work surfaces of biological safety cabinets and other containment equipment are decontaminated when work with infectious materials is finished. Plastic-backed paper toweling used on nonperforated work surfaces within biological safety cabinets facilitates clean-up.

8. An insect and rodent control program is in effect.

9. Laboratory clothing that protects street clothing (e.g., solid-front or wrap-around gowns, scrub suits, coveralls) is worn in the laboratory. Laboratory clothing is not worn outside the laboratory, and it is decontaminated before being laundered.

10. Special care is taken to avoid skin contamination with infectious materials; gloves are worn when handling infected animals and when skin contact with infectious materials is unavoidable.

11. Molded surgical masks or respirators are worn in rooms containing infected animals.

12. Animals and plants not related to the work being conducted are not permitted in the laboratory.

13. All waste from laboratories and animal rooms is appropriately decontaminated before being disposed of.

14. Vacuum lines are protected with high-efficiency particulate air (HEPA) filters and liquid disinfectant traps.

15. Hypodermic needles and syringes are used only for parenteral injection and aspiration of fluids from laboratory animals and diaphragm bottles. Only needle-locking syringes or disposable syringe-needle units (i.e., the needle is integral to the syringe) are used for the injection or aspiration of infectious fluids. Extreme caution is used when handling needles and syringes to avoid autoinoculation and the generation of aerosols during use and disposal. A needle should not be bent, sheared, replaced in the sheath or guard, or removed from the syringe following use. The needle and syringe should be promptly placed in a puncture-resistant container and decontaminated, preferably by autoclaving, before being discarded or reused.

16. Spills and accidents that result in overt or potential exposures to infectious material are immediately reported to the laboratory director. Appropriate medical evaluation, surveillance, and treatment are provided, and written records are maintained.

17. Baseline serum samples for all laboratory and other at-risk personnel are collected and stored. Additional serum specimens may be collected periodically, depending on the agents handled or the function of the laboratory.

18. A biosafety manual is prepared or adopted. Personnel are advised of special hazards and are required to read instructions on practices and procedures and to follow them.

C. *Containment equipment*

Biological safety cabinets (Class I, II, or III) or other appropriate combinations of personal-protection or physical-containment devices (e.g., special protective clothing, masks, gloves, respirators, centrifuge safety cups, sealed centrifuge rotors, and containment caging for animals) are used for all activities with infectious materials that pose a threat of aerosol exposure. These include: manipulation of cultures and of clinical or environmental material that may be a source of infectious aerosols; the aerosol challenge of experimental animals; harvesting of tissues or fluids from infected animals and embryonated eggs; and necropsy of infected animals.

D. *Laboratory facilities*

1. The laboratory is separated from areas that are open to unrestricted traffic flow within the building. Passage through two sets of doors is the basic requirement for entry into the laboratory from access corridors or other contiguous areas. Physical separation of the high-containment laboratory from access corridors or other laboratories or activities may also be provided by a double-doored clothes-change room (showers may be included), airlock, or other access facility that requires passing through two sets of doors before entering the laboratory.

2. The interior surfaces of walls, floors, and ceilings are water resistant so that they can be easily cleaned. Penetrations in these surfaces are sealed or

capable of being sealed to facilitate decontaminating the area.

3. Bench tops are impervious to water and resistant to acids, alkalis, organic solvents, and moderate heat.

4. Laboratory furniture is sturdy, and spaces between benches, cabinets, and equipment are accessible for cleaning.

5. Each laboratory contains a sink for washing hands. The sink is foot-, elbow-, or automatically operated and is located near the laboratory exit door.

6. Windows in the laboratory are closed and sealed.

7. Access doors to the laboratory or containment module are self-closing.

8. An autoclave for decontaminating laboratory wastes is available, preferably within the laboratory.

9. A ducted exhaust-air ventilation system is provided. This system creates directional airflow that draws air into the laboratory through the entry area. The exhaust air is not recirculated to any other area of the building, is discharged to the outside, and is dispersed away from occupied areas and air intakes. Personnel must verify that the direction of the airflow is proper (i.e., into the laboratory). The exhaust air from the laboratory room can be discharged to the outside without being filtered or otherwise treated.

10. The HEPA-filtered exhaust air from Class I or Class II biological safety cabinets is discharged directly to the outside or through the building exhaust system. Exhaust air from Class I or II biological safety cabinets may be recirculated within the laboratory if the cabinet is tested and certified at least every 12 months. If the HEPA-filtered exhaust air from Class I or II biological safety cabinets is to be discharged to the outside through the building exhaust system, it is connected to this system in a manner (e.g., thimble-unit connection) that avoids any interference with the air balance of the cabinets or building exhaust system.

VERTEBRATE ANIMAL BIOSAFETY LEVEL CRITERIA

Animal Biosafety Level 2

A. *Standard practices*

1. Doors to animal rooms open inward, are self-closing, and are kept closed when infected animals are present.

2. Work surfaces are decontaminated after use or spills of viable materials.

3. Eating, drinking, smoking, and storing of food for human use are not permitted in animal rooms.

4. Personnel wash their hands after handling cultures and animals and before leaving the animal room.

5. All procedures are carefully performed to minimize the creation of aerosols.

6. An insect and rodent control program is in effect.

B. *Special practices*

1. Cages are decontaminated, preferably by autoclaving, before being cleaned and washed.

2. Surgical-type masks are worn by all personnel entering animal rooms housing nonhuman primates.

3. Laboratory coats, gowns, or uniforms are worn while in the animal room. This protective clothing is removed before leaving the animal facility.

4. The laboratory or animal-facility director limits access to the animal room only to personnel who have been advised of the potential hazard and who need to enter the room for program or service purposes when work is in progress. In general, persons who may be at increased risk of acquiring infection or for whom infection might be unusually hazardous are not allowed in the animal room.

5. The laboratory or animal-facility director establishes policies and procedures whereby only persons who have been advised of the potential hazard and who meet any specific requirements (e.g., vaccination) may enter the animal room.

6. When an infectious agent in use in the animal room requires special-entry provisions (e.g., vaccination), a hazard warning sign (incorporating the universal biohazard symbol) is posted on the access door to the animal room. The hazard warning sign identifies the infectious agent, lists the name and telephone number of the animal-facility supervisor or other responsible person(s), and indicates the special requirement(s) for entering the animal room.

7. Special care is taken to avoid contaminating skin with infectious material; gloves should be worn when handling infected animals and when skin contact with infectious materials is unavoidable.

8. All waste from the animal room is appropriately decontaminated—preferably by autoclaving—

before being disposed of. Infected animal carcasses are incinerated after being transported from the animal room in leakproof, covered containers.

9. Hypodermic needles and syringes are used only for the parenteral injection or aspiration of fluids from laboratory animals and diaphragm bottles. Only needle-locking syringes or disposable syringe-needle units (i.e., the needle is integral to the syringe) are used for the injection or aspiration of infectious fluids. A needle should not be bent, sheared, replaced in the sheath or guard, or removed from the syringe following use. The needle and syringe should be promptly placed in a puncture-resistant container and decontaminated, preferably by autoclaving, before being discarded or reused.

10. If floor drains are provided, the drain taps are always filled with water or a suitable disinfectant.

11. When appropriate, considering the agents handled, baseline serum samples from animal-care and other at-risk personnel are collected and stored. Additional serum samples may be collected periodically, depending on the agents handled or the function of the facility.

C. *Containment equipment*

Biological safety cabinets, other physical-containment devices, and/or personal-protection devices (e.g., respirators, face shields) are used when procedures with a high potential for creating aerosols are conducted. These include necropsy of infected animals, harvesting of infected tissues or fluids from animals or eggs, intranasal inoculation of animals, and manipulation of high concentrations or large volumes of infectious materials.

D. *Animal facilities*

1. The animal facility is designed and constructed to facilitate cleaning and housekeeping.

2. A sink for washing hands is available in the room that houses infected animals.

3. If the animal facility has windows that open, they are fitted with fly screens.

4. It is recommended, but not required, that the direction of airflow in the animal facility is inward and that exhaust air is discharged to the outside without being recirculated to other rooms.

5. An autoclave that can be used for decontaminating infectious laboratory waste is available in the same building that contains the animal facility.

Animal Biosafety Level 3

A. *Standard practices*

1. Doors to animal rooms open inward, are self-closing, and are kept closed when work with infected animals is in progress.

2. Work surfaces are decontaminated after use or after spills of viable materials.

3. Eating, drinking, smoking, and storing of food for human use are not permitted in the animal room.

4. Personnel wash their hands after handling cultures or animals and before leaving the laboratory.

5. All procedures are carefully performed to minimize the creation of aerosols.

6. An insect and rodent control program is in effect.

B. *Special practices*

1. Cages are autoclaved before bedding is removed and before they are cleaned and washed.

2. Surgical-type masks or other respiratory protection devices (e.g., respirators) are worn by personnel entering rooms that house animals infected with agents assigned to Biosafety Level 3.

3. Wrap-around or solid-front gowns or uniforms are worn by personnel entering the animal room. Front-button laboratory coats are unsuitable. Protective gowns must remain in the animal room and must be decontaminated before being laundered.

4. The laboratory director or other responsible person limits access to the animal room only to personnel who have been advised of the potential hazard and who need to enter the room for program or service purposes when infected animals are present. In general, persons who may be at increased risk of acquiring infection or for whom infection might be unusually hazardous are not allowed in the animal room.

5. The laboratory director or other responsible person establishes policies and procedures whereby only persons who have been advised of the potential hazard and meet any specific requirements (e.g., vaccination) may enter the animal room.

6. Hazard warning signs (incorporating the universal biohazard warning symbol) are posted on access doors to animal rooms containing animals infected with agents assigned to Biosafety Level 3 are present. The hazard warning sign should identify the agent(s) in use, list the name and telephone number of the animal room supervisor or other responsible person(s), and indicate any special conditions of entry into the animal room (e.g., the need for vaccinations or respirators).

7. Personnel wear gloves when handling infected animals. Gloves are removed aseptically and autoclaved with other animal room waste before being disposed of or reused.

8. All wastes from the animal room are autoclaved before being disposed of. All animal carcasses are incinerated. Dead animals are transported from the animal room to the incinerator in leakproof, covered containers.

9. Hypodermic needles and syringes are used only for gavage or parenteral injection or aspiration of fluids from laboratory animals and diaphragm bottles. Only needle-locking syringes or disposable syringe-needle units (i.e., the needle is integral to the syringe) are used. A needle should not be bent, sheared, replaced in the sheath or guard, or removed from the syringe following use. The needle and syringe should be promptly placed in a puncture-resistant container and decontaminated, preferably by autoclaving, before being discarded or reused. When possible, cannulas should be used instead of sharp needles (e.g., for gavage).

10. If floor drains are provided, the drain traps are always filled with water or a suitable disinfectant.

11. If vacuum lines are provided, they are protected with HEPA filters and liquid disinfectant traps.

12. Boots, shoe covers, or other protective footwear and disinfectant footbaths are available and used when indicated.

C. *Containment equipment*

1. Personal-protection clothing and equipment and/or other physical-containment devices are used for all procedures and manipulations of infectious materials or infected animals.

2. The risk of infectious aerosols from infected animals or their bedding can be reduced if animals are housed in partial-containment caging systems, such as open cages placed in ventilated enclosures (e.g., laminar-flow cabinets), solid-wall and -bottom cages covered by filter bonnets, or other equivalent primary containment systems.

D. *Animal facilities*

1. The animal facility is designed and constructed to facilitate cleaning and housekeeping and is separated from areas that are open to unrestricted personnel traffic within the building. Passage through two sets of doors is the basic requirement for entry into the animal room from access corridors or other contiguous areas. Physical separation of the animal room from access corridors or from other activities may also be provided by a double-doored clothes change room (showers may be included), airlock, or other access facility that requires passage through two sets of doors before entering the animal room.

2. The interior surfaces of walls, floors, and ceilings are water resistant so that they can be cleaned easily. Penetrations in these surfaces are sealed or capable of being sealed to facilitate fumigation or space decontamination.

3. A foot-, elbow-, or automatically operated sink for handwashing is provided near each animal-room exit door.

4. Windows in the animal room are closed and sealed.

5. Animal room doors are self-closing and are kept closed when infected animals are present.

6. An autoclave for decontaminating wastes is available, preferably within the animal room. Materials to be autoclaved outside the animal room are transported in a covered, leakproof container.

7. An exhaust-air ventilation system is provided. This system creates directional airflow that draws air into the animal room through the entry area. The building exhaust can be used for this purpose if the exhaust air is not recirculated to any other area of the building, is discharged to the outside, and is dispersed away from occupied areas and air intakes. Personnel must verify that the direction of the airflow is proper (i.e., into the animal room). The exhaust air from the animal room that does not pass through biological safety cabinets or other primary containment equipment can be discharged to the outside without being filtered or otherwise treated.

8. The HEPA-filtered exhaust air from Class I or Class II biological safety cabinets or other primary

containment devices is discharged directly to the outside or through the building's exhaust system. Exhaust air from these primary containment devices may be recirculated within the animal room if the cabinet is tested and certified at least every 12 months. If the HEPA-filtered exhaust air from Class I or Class II biological safety cabinets is discharged to the outside through the building exhaust system, it is connected to this system in a manner (e.g., thimble-unit connection) that avoids any interference with the air balance of the cabinets or building exhaust system.

ADDENDUM 2: CDC CAUTIONARY NOTICE

CDC cautionary notice for all human-serum-derived reagents used as controls:

> WARNING: Because no test method can offer complete assurance that laboratory specimens do not contain HIV, hepatitis B virus, or other infectious agents, this specimen should be handled at the BSL 2 as recommended for any potentially infectious human serum or blood specimen in the CDC-NIH manual, *Biosafety in Microbiological and Biomedical Laboratories,* 1984, pages 11-13.

If additional statements describing the results of any heat treatment or serologic procedure(s) already performed on the human-serum reagent or control are used in conjunction with the above cautionary notice, these statements should be worded so as not to diminish the impact of the warning that emphasizes the need for universal precautions.

REFERENCES

1. Richardson JH, Barkley WE, eds. Biosafety in microbiological and biomedical laboratories, 1984. Washington, DC: Public Health Service, 1984; DHHS publication no. (CDC)84-8395.

2. CDC. Human T-lymphotropic virus type III/ lymphadenopathy-associated virus agent summary statement. MMWR 1986;35:540-2, 547-9.

3. CDC. Recommendations for prevention of HIV transmission in health-care settings. MMWR 1987;36(suppl 2):3S-18S.

4. Isenberg HD, Washington JA, Balows A, Sonnenwirth AC. Collection, handling and processing of specimens. In: Lennette EH, Balows A, Hausler WJ, Shadomy HJ, eds. Manual of clinical microbiology, 4th ed. Washington, DC: American Society for Microbiology, 1985:73-98.

5. CDC. Acquired immune deficiency syndrome (AIDS): precautions for clinical and laboratory workers. MMWR 1982;32:577-80.

6. CDC. Recommendations for preventing transmission of infection with human T-lymphotropic virus type III/lymphadenopathy-associated virus in the workplace. MMWR 1985;34:681.

7. CDC. Recommendations for preventing transmission of infection with human T-lymphotropic virus type III/lymphadenopathy-associated virus during invasive procedures. MMWR 1986;35:221-3.

8. CDC. Recommendations for preventing transmission of infection with human T-lymphotropic virus type III/lymphadenopathy-associated virus in the workplace. MMWR 1985;34:682-6, 691-5.

9. U.S. Department of Health and Human Services, Public Health Service. Joint advisory notice: HBV/HIV. Federal Register 1987;52 (October 30): 41818-24.

10. CDC. Update: evaluation of human T-lymphotropic virus type III/lymphadenopathy-associated virus infection in health-care personnel—United States. MMWR 1985;34:575-8.

11. CDC. Human immunodeficiency virus infections in health-care workers exposed to blood of infected patients. MMWR 1987;36:285-9.

12. Anonymous. Needlestick transmission of HTLV-III from a patient infected in Africa. Lancet 1984;2:1376-7.

13. Stricof RL, Morse DL. HTLV-III/LAV seroconversion following a deep intramuscular needlestick injury. New Engl J Med 1986; 314:1115.

14. McCray E, Cooperative Needlestick Study Group. Occupational risk of the acquired immunodeficiency syndrome among health-care workers. New Engl J Med 1986;314:1127-32.

15. Weiss SH, Saxinger WC, Richtman D, et al. HTLV-III infection among health-care workers: association with needlestick injuries. JAMA 1985;254:2089-93.

16. Henderson DK, Saah AJ, Zak BJ, et al. Risk of nosocomial infection with human T-cell lymphotropic virus type III/lymphadenopathy-associated virus in a large cohort of intensively exposed health care workers. Ann Intern Med 1986;104:644-7.

17. Gerberding JL, Bryant-Le Blanc CE, Nelson K, et al. Risk of transmitting the human immunodeficiency virus, cytomegalovirus, and hepatitis B virus to health-care workers exposed to patients with AIDS and AIDS-related conditions. J Infect Dis 1987;156:1-8.

18. Weiss SH, Goedert JJ, Gartner S, et al. Risk of human immunodeficiency virus (HIV-1) infection among laboratory workers. Science 1988;239:68-71.

19. U.S. Department of Health and Human Services, Public Health Service. Biosafety guidelines for use of HTLV-III and related viruses. Federal Register 1984 (October 16);49:40556.

20. U.S. Environmental Protection Agency. EPA guide for infectious waste management. Washington, DC: U.S. Environmental Protection Agency, 1986; Publication no. EPA/530-5W-86-014.

21. Favero MS. Sterilization, disinfection and antisepsis in the hospital. In: Lennette EH, Balows A, Hausler WJ, Shadomy HJ, eds. Manual of clinical microbiology, 4th ed. Washington, DC: American Society for Microbiology, 1985:129-37.

22. Martin LS, McDougal JS, Loskoski SL. Disinfection and inactivation of the human T lymphotropic virus type III/lymphadenopathy-associated virus. J Infect Dis 1985;152:400-3.

23. Resnick L, Veren K, Salahuddin, SZ, Tondreau S, Markham PD. Stability and inactivation of HTLV-III/LAV under clinical and laboratory environments. JAMA 1986;255:1887-91.

24. Favero MS, Petersen NJ, Bond WW. Transmission and control of laboratory-acquired hepatitis infection. In: Miller BM, Groschel DHM, Richardson JH, et al., eds. Laboratory safety: principles and practice. Washington, DC: American Society for Microbiology, 1986:49-58.

25. CDC. Public Health Service guidelines for counseling and antibody testing to prevent HIV infections and AIDS. MMWR 1987;36:509-15.

26. Ronalds CJ, Grint PCA, Kangro HD. Disinfection and inactivation of HTLV-III/LAV (Letter). J Infect Dis 1986;153:996.

27. Evans RP, Shanson DC. Effect of heat on serologic tests for hepatitis B and syphilis and on aminoglycoside assays (Letter). Lancet 1985;1:1458.

28. Van den Akker R, Hekker AC, Osterhaus ADME. Heat inactivation of serum may interfere with HTLV-III/LAV serology (Letter). Lancet 1985;2:672.

29. Mortimer PP, Parry JV, Mortimer JY. Which anti-HTLV III/LAV assays for screening and confirmatory testing? Lancet 1985;2:873-7.

30. Jungkind DL, DiRenzo SA, Young SJ. Effect of using heat-inactivated serum with the Abbott human T-cell lymphotropic virus type III antibody test. J Clin Microbiol 1986;23:381-2.

31. Goldie DJ, McConnell AA, Cooke PR. Heat treatment of whole blood and serum before chemical analysis (Letter). Lancet 1985;1:1161.

32. Lai L, Ball G, Stevens J, Shanson D. Effect of heat treatment of plasma and serum on biochemical indices (Letter). Lancet 1985;1:1457-8.

33. CDC. Additional recommendations to reduce sexual and drug abuse-related transmission of human T-lymphotropic virus type III/lymphadenopathy-associated virus. MMWR 1986; 35:152-5.

34. CDC. Revision of the case definition of acquired immunodeficiency syndrome for national reporting—United States. MMWR 1985; 34:373-5.

35. CDC. Diagnosis and management of mycobacterial infection and disease in persons with human T-lymphotropic virus type III/lymphadenopathy-associated virus infection. MMWR 1986;35:448-52.

36. CDC. Revision of the CDC surveillance case definition for acquired immunodeficiency syndrome. MMWR 1987;36(suppl 1):1S-15S.

Appendix C

Recommendations for Prevention of HIV Transmission in Health-Care Settings

CONTENTS

I. Recommendations for Prevention of HIV Transmission in Health-Care Settings*

---

* Reprinted from *Morbidity and Mortality Weekly Report,* 1987; 36 (2S):3S-18S.

NOTE: Tables and page numbers in these reprints have been renumbered to avoid confusion.

II. Update: Universal Precautions for Prevention of Transmission of Human Immunodeficiency Virus, Hepatitis B Virus, and Other Bloodborne Pathogens in Health-Care Settings**

I. RECOMMENDATIONS FOR PREVENTION OF HIV TRANSMISSION IN HEALTH-CARE SETTINGS

INTRODUCTION

Human immunodeficiency virus (HIV), the virus that causes acquired immunodeficiency syndrome (AIDS), is transmitted through sexual contact and exposure to infected blood or blood components and perinatally from mother to neonate. HIV has been isolated from blood, semen, vaginal secretions, saliva, tears, breast milk, cerebrospinal fluid, amniotic fluid, and urine and is likely to be isolated from other body fluids, secretions, and excretions. However, epidemiologic evidence has implicated only blood, semen, vaginal secretions, and possibly breast milk in transmission.

The increasing prevalence of HIV increases the risk that health-care workers will be exposed to blood from patients infected with HIV, especially when blood and body-fluid precautions are not followed for all patients. Thus, this document emphasizes the need for health-care workers to consider *all* patients as potentially infected with HIV and/or other bloodborne pathogens and to adhere rigorously to infection-control precautions for minimizing the risk of exposure to blood and body fluids of all patients.

The recommendations contained in this document consolidate and update CDC recommendations published earlier for preventing HIV transmission in health-care settings: precautions for clinical and laboratory staffs (1) and precautions for health-care workers and allied professionals (2); recommendations for preventing HIV transmission in the workplace (3) and during invasive procedures (4); recommendations for preventing possible transmission of HIV from tears (5); and recommendations for providing dialysis treatment for HIV-infected patients (6). These recommendations also update portions of the "Guideline for Isolation Precautions in Hospitals" (7) and reemphasize some of the recommendations contained in "Infection Control Practices for Dentistry" (8). The recommendations contained in this document have been developed for use in health-care settings and emphasize the need to treat blood and other body fluids from *all* patients as potentially infective. These same prudent precautions also should be taken in other settings in which persons may be exposed to blood or other body fluids.

DEFINITION OF HEALTH-CARE WORKERS

Health-care workers are defined as persons, including students and trainees, whose activities involve contact with patients or with blood or other body fluids from patients in a health-care setting.

** Reprinted from *Morbidity and Mortality Weekly Report,* 1988; 37(24):377+.

HEALTH-CARE WORKERS WITH AIDS

As of July 10, 1987, a total of 1,875 (5.8%) of 32,395 adults with AIDS, who had been reported to the CDC national surveillance system and for whom occupational information was available, reported being employed in a health-care or clinical laboratory setting. In comparison, 6.8 million persons—representing 5.6% of the U.S. labor force—were employed in health services. Of the health-care workers with AIDS, 95% have been reported to exhibit high-risk behavior; for the remaining 5%, the means of HIV acquisition was undetermined. Health-care workers with AIDS were significantly more likely than other workers to have an undetermined risk (5% versus 3%, respectively). For both health-care workers and non-health-care workers with AIDS, the proportion with an undetermined risk has not increased since 1982.

AIDS patients initially reported as not belonging to recognized risk groups are investigated by state and local health departments to determine whether possible risk factors exist. Of all health-care workers with AIDS reported to CDC who were initially characterized as not having an identified risk and for whom follow-up information was available, 66% have been reclassified because risk factors were identified or because the patient was found not to meet the surveillance case definition for AIDS.

Of the 87 health-care workers currently categorized as having no identifiable risk, information is incomplete on 16 (18%) because of death or refusal to be interviewed; 38 (44%) are still being investigated.

The remaining 33 (38%) health-care workers were interviewed or had other follow-up information available. The occupations of these 33 were as follows:

nine nursing assistants	(27%);
seven housekeeping or maintenance workers	(21%);
five physicians, three of whom were surgeons	(15%);
three nurses	(9%);
three clinical laboratory technicians	(9%);
one dentist	(3%);
one therapist	(3%); and
four others who did not have contact with patients	(12%).

Although 15 of these 33 health-care workers reported parenteral and/or other non-needlestick exposure to blood or body fluids from patients in the 10 years preceding their diagnosis of AIDS, none of these exposures involved a patient with AIDS or known HIV infection.

RISK TO HEALTH-CARE WORKERS OF ACQUIRING HIV IN HEALTH-CARE SETTINGS

Health-care workers with documented percutaneous or mucous-membrane exposures to blood or body fluids of HIV-infected patients have been prospectively evaluated to determine the risk of infection after such exposures. As of June 30, 1987, 883 health-care workers have been tested for antibody to HIV in an ongoing surveillance project conducted by CDC (9). Of these, 708 (80%) had percutaneous exposures to blood, and 175 (20%) had a mucous membrane or open wound contaminated by blood or body fluid. Of 396 health-care workers, each of whom had only a convalescent-phase serum sample obtained and tested ≥ 90 days postexposure, one—for whom heterosexual transmission could not be ruled out—was seropositive for HIV antibody. For 425 additional health-care workers, both acute- and convalescent-phase serum samples were obtained and tested; none of 74 health-care workers with nonpercutaneous exposures seroconverted, and three (0.9%) of 351 with percutaneous exposures seroconverted. None of these three health-care workers had other documented risk factors for infection.

Two other prospective studies to assess the risk of nosocomial acquisition of HIV infection for health-care workers are ongoing in the United States. As of April 30, 1987, 332 health-care workers with a total of 453 needlestick or mucous-membrane exposures to the blood or other body fluids of HIV-infected patients were tested for HIV antibody at the National Institutes of Health (10). These exposed workers included 103 with needlestick injuries and 229 with mucous-membrane exposures; none had seroconverted.

A similar study at the University of California of 129 health-care workers with documented needlestick injuries or mucous-membrane exposures to blood or other body fluids from patients with HIV infection has not identified any seroconversions (11). Results of a prospective study in the United Kingdom identified no evidence of transmission among 150 health-care workers with parenteral or mucous-membrane exposures to blood or other body fluids, secretions, or excretions from patients with HIV infection (12).

In addition to health-care workers enrolled in prospective studies, eight persons who provided care to infected patients and denied other risk factors have been reported to have acquired HIV infection. Three of these health-care workers had needlestick exposures to blood from infected patients (13-15). Two were persons who provided nursing care to infected persons; although neither sustained a needlestick, both had extensive contact with blood or other body fluids, and neither observed recommended barrier precautions (16,17). The other three were health-care workers with non-needlestick exposures to blood from infected patients (18). Although the exact route of transmission for these last three infections is not known, all three persons had direct contact of their skin with blood from infected patients, all had skin lesions that may have been contaminated by blood, and one also had a mucous-membrane exposure.

A total of 1,231 dentists and hygienists, many of whom practiced in areas with many AIDS cases, participated in a study to determine the prevalence of antibody to HIV; one dentist (0.1%) had HIV antibody. Although no exposure to a known HIV-infected person could be documented, epidemiologic investigation did not identify any other risk factor for infection. The infected dentist, who also had a history of sustaining needlestick injuries and trauma to his hands, did not routinely wear gloves when providing dental care (19).

PRECAUTIONS TO PREVENT TRANSMISSION OF HIV

Universal Precautions

Since medical history and examination cannot reliably identify all patients infected with HIV or other blood-borne pathogens, blood and body-fluid precautions should be consistently used for *all* patients. This approach, previously recommended by CDC (3,4), and referred to as "universal blood and body-fluid precautions" or "universal precautions," should be used in the care of *all* patients, especially including those in emergency-care settings in which the risk of blood exposure is increased and the infection status of the patient is usually unknown (20).

1. All health-care workers should routinely use appropriate barrier precautions to prevent skin and mucous-membrane exposure when contact with blood or other body fluids of any patient is anticipated. Gloves should be worn for touching blood and body fluids, mucous membranes, or non-intact skin of all patients, for handling items or surfaces soiled with blood or body fluids, and for performing venipuncture and other vascular access procedures. Gloves should be changed after contact with each patient. Masks and protective eye wear or face shields should be worn during procedures that are likely to generate droplets of blood or other body fluids to prevent exposure of mucous membranes of the mouth, nose, and eyes. Gowns or aprons should be worn during procedures that are likely to generate splashes of blood or other body fluids.

2. Hands and other skin surfaces should be washed immediately and thoroughly if contaminated with blood or other body fluids. Hands should be washed immediately after gloves are removed.

3. All health-care workers should take precautions to prevent injuries caused by needles, scalpels, and other sharp instruments or devices during procedures; when cleaning used instruments; during disposal of used needles; and when handling sharp instruments after procedures. To prevent needlestick injuries, needles should not be recapped, purposely bent or broken by hand, removed from disposable syringes, or otherwise manipulated by hand. After they are used, disposable syringes and needles, scalpel blades, and other sharp items should be placed in puncture-resistant containers for disposal; the puncture-resistant containers should be located as close as practical to the use area. Large-bore reusable needles should be placed in a puncture-resistant container for transport to the reprocessing area.

4. Although saliva has not been implicated in HIV transmission, to minimize the need for emergency mouth-to-mouth resuscitation, mouthpieces, resuscitation bags, or other ventilation devices should be available for use in areas in which the need for resuscitation is predictable.

5. Health-care workers who have exudative lesions or weeping dermatitis should refrain from all direct patient care and from handling patient-care equipment until the condition resolves.

6. Pregnant health-care workers are not known to be at greater risk of contracting HIV infection than

health-care workers who are not pregnant; however, if a health-care worker develops HIV infection during pregnancy, the infant is at risk of infection resulting from perinatal transmission. Because of this risk, pregnant health-care workers should be especially familiar with and strictly adhere to precautions to minimize the risk of HIV transmission.

Implementation of universal blood and body-fluid precautions for *all* patients eliminates the need for use of the isolation category of "Blood and Body Fluid Precautions" previously recommended by CDC (7) for patients known or suspected to be infected with blood-borne pathogens. Isolation precautions (e.g., enteric, AFB (7)) should be used as necessary if associated conditions, such as infectious diarrhea or tuberculosis, are diagnosed or suspected.

Precautions for Invasive Procedures

In this document, an invasive procedure is defined as surgical entry into tissues, cavities, or organs or repair of major traumatic injuries 1) in an operating or delivery room, emergency department, or outpatient setting, including both physicians' and dentists' offices; 2) cardiac catheterization and angiographic procedures; 3) a vaginal or cesarean delivery or other invasive obstetric procedure during which bleeding may occur; or 4) the manipulation, cutting, or removal of any oral or perioral tissues, including tooth structure, during which bleeding occurs or the potential for bleeding exists. The universal blood and body-fluid precautions listed above, combined with the precautions listed below, should be the minimum precautions for *all* such invasive procedures.

1. All health-care workers who participate in invasive procedures must routinely use appropriate barrier precautions to prevent skin and mucous-membrane contact with blood and other body fluids of all patients. Gloves and surgical masks must be worn for all invasive procedures. Protective eye wear or face shields should be worn for procedures that commonly result in the generation of droplets, splashing of blood or other body fluids, or the generation of bone chips. Gowns or aprons made of materials that provide an effective barrier should be worn during invasive procedures that are likely to result in the splashing of blood or other body fluids. All health-care workers who perform or assist in vaginal or cesarean deliveries should wear gloves and gowns when handling the placenta or the infant until blood and amniotic fluid have been removed from the infant's skin and should wear gloves during post-delivery care of the umbilical cord.

2. If a glove is torn or a needlestick or other injury occurs, the glove should be removed and a new glove used as promptly as patient safety permits; the needle or instrument involved in the incident should also be removed from the sterile field.

Precautions for Dentistry***

Blood, saliva, and gingival fluid from *all* dental patients should be considered infective. Special emphasis should be placed on the following precautions for preventing transmission of blood-borne pathogens in dental practice in both institutional and non-institutional settings.

1. In addition to wearing gloves for contact with oral mucous membranes of all patients, all dental workers should wear surgical masks and protective eye wear or chin-length plastic face shields during dental procedures in which splashing or spattering of blood, saliva, or gingival fluids is likely. Rubber dams, high-speed evacuation, and proper patient positioning, when appropriate, should be utilized to minimize generation of droplets and spatter.

2. Handpieces should be sterilized after use with each patient, since blood, saliva, or gingival fluid of patients may be aspirated into the handpiece or waterline. Handpieces that cannot be sterilized should at least be flushed, the outside surface cleaned and wiped with a suitable chemical germicide, and then rinsed. Handpieces should be flushed at the beginning of the day and after use with each patient. Manufacturers' recommendations should be followed for use and maintenance of waterlines and check valves and for flushing of handpieces. The same precautions should be used for ultrasonic scalers and air/water syringes.

3. Blood and saliva should be thoroughly and carefully cleaned from material that has been used in

***General infection-control precautions are more specifically addressed in previous recommendations for infection-control practices for dentistry (8).

the mouth (e.g., impression materials, bite registration), especially before polishing and grinding intra-oral devices. Contaminated materials, impressions, and intra-oral devices should also be cleaned and disinfected before being handled in the dental laboratory and before they are placed in the patient's mouth. Because of the increasing variety of dental materials used intra-orally, dental workers should consult with manufacturers as to the stability of specific materials when using disinfection procedures.

4. Dental equipment and surfaces that are difficult to disinfect (e.g., light handles or X-ray-unit heads) and that may become contaminated should be wrapped with impervious-backed paper, aluminum foil, or clear plastic wrap. The coverings should be removed and discarded, and clean coverings should be put in place after use with each patient.

Precautions for Autopsies or Morticians' Services

In addition to the universal blood and body-fluid precautions listed above, the following precautions should be used by persons performing post-mortem procedures:

1. All persons performing or assisting in post-mortem procedures should wear gloves, masks, protective eye wear, gowns, and waterproof aprons.
2. Instruments and surfaces contaminated during post-mortem procedures should be decontaminated with an appropriate chemical germicide.

Precautions for Dialysis

Patients with end-stage renal disease who are undergoing maintenance dialysis and who have HIV infection can be dialyzed in hospital-based or free-standing dialysis units using conventional infection-control precautions (21). Universal blood and body-fluid precautions should be used when dialyzing all patients.

Strategies for disinfecting the dialysis fluid pathways of the hemodialysis machine are targeted to control bacterial contamination and generally consist of using 500-750 parts per million (ppm) of sodium hypochlorite (household bleach) for 30-40 minutes or 1.5%-2.0% formaldehyde overnight. In addition, several chemical germicides formulated to disinfect dialysis machines are commercially available. None of these protocols or procedures need to be changed for dialyzing patients infected with HIV.

Patients infected with HIV can be dialyzed by either hemodialysis or peritoneal dialysis and do not need to be isolated from other patients. The type of dialysis treatment (i.e., hemodialysis or peritoneal dialysis) should be based on the needs of the patient. The dialyzer may be discarded after each use. Alternatively, centers that reuse dialyzers—i.e., a specific single-use dialyzer is issued to a specific patient, removed, cleaned, disinfected, and reused several times on the same patient only—may include HIV-infected patients in the dialyzer-reuse program. An individual dialyzer must never be used on more than one patient.

Precautions for Laboratories****

Blood and other body fluids from all patients should be considered infective. To supplement the universal blood and body-fluid precautions listed above, the following precautions are recommended for health care personnel:

1. All specimens of blood and body fluids should be put in a well-constructed container with a secure lid to prevent leaking during transport. Care should be taken when collecting each specimen to avoid contaminating the outside of the container and of the laboratory form accompanying the specimen.
2. All persons processing blood and body-fluid specimens (e.g., removing tops from vacuum tubes) should wear gloves. Masks and protective eye wear should be worn if mucous-membrane contact with blood or body fluids is anticipated. Gloves should be changed and hands washed after completion of specimen processing.
3. For routine procedures, such as histologic and pathologic studies or microbiologic culturing, a biological safety cabinet is not necessary. However, biological safety cabinets (Class I or II) should be used whenever procedures are conducted that have a high potential for generating droplets. These include activities such as blending, sonicating, and vigorous mixing.

****Additional precautions for research and industrial laboratories are addressed elsewhere (22,23).

4. Mechanical pipetting devices should be used for manipulating all liquids in the laboratory. Mouth pipetting must not be done.

5. Use of needles and syringes should be limited to situations in which there is no alternative, and the recommendations for preventing injuries with needles outlined under universal precautions should be followed.

6. Laboratory work surfaces should be decontaminated with an appropriate chemical germicide after a spill of blood or other body fluids and when work activities are completed.

7. Contaminated materials used in laboratory tests should be decontaminated before reprocessing or be placed in bags and disposed of in accordance with institutional policies for disposal of infective waste (24).

8. Scientific equipment that has been contaminated with blood or other body fluids should be decontaminated and cleaned before being repaired in the laboratory or transported to the manufacturer.

9. All persons should wash their hands after completing laboratory activities and should remove protective clothing before leaving the laboratory.

Implementation of universal blood and body-fluid precautions for *all* patients eliminates the need for warning labels on specimens since blood and other body fluids from all patients should be considered infective.

ENVIRONMENTAL CONSIDERATIONS FOR HIV TRANSMISSION

No environmentally mediated mode of HIV transmission has been documented. Nevertheless, the precautions described below should be taken routinely in the care of all patients.

Sterilization and Disinfection

Standard sterilization and disinfection procedures for patient-care equipment currently recommended for use (25,26) in a variety of health-care settings—including hospitals, medical and dental clinics and offices, hemodialysis centers, emergency-care facilities, and long-term nursing-care facilities—are adequate to sterilize or disinfect instruments, devices, or other items contaminated with blood or other body fluids from persons infected with blood-borne pathogens including HIV (21,23).

Instruments or devices that enter sterile tissue or the vascular system of any patient or through which blood flows should be sterilized before reuse. Devices or items that contact intact mucous membranes should be sterilized or receive high-level disinfection, a procedure that kills vegetative organisms and viruses but not necessarily large numbers of bacterial spores. Chemical germicides that are registered with the U.S. Environmental Protection Agency (EPA) as "sterilants" may be used either for sterilization or for high-level disinfection depending on contact time.

Contact lenses used in trial fittings should be disinfected after each fitting by using a hydrogen peroxide contact lens disinfecting system or, if compatible, with heat (78°C-80°C (172.4°F-176.0°F)) for 10 minutes.

Medical devices or instruments that require sterilization or disinfection should be thoroughly cleaned before being exposed to the germicide, and the manufacturer's instructions for the use of the germicide should be followed. Further, it is important that the manufacturer's specifications for compatibility of the medical device with chemical germicides be closely followed. Information on specific label claims of commercial germicides can be obtained by writing to Disinfectants Branch, Office of Pesticides, Environmental Protection Agency, 401 M Street, SW, Washington, D.C. 20460.

Studies have shown that HIV is inactivated rapidly after being exposed to commonly used chemical germicides at concentrations that are much lower than used in practice (27-30). Embalming fluids are similar to the types of chemical germicides that have been tested and found to completely inactivate HIV. In addition to commercially available chemical germicides, a solution of sodium hypochlorite (household bleach) prepared daily is an inexpensive and effective germicide. Concentrations ranging from approximately 500 ppm (1:100 dilution of household bleach) sodium hypochlorite to 5,000 ppm (1:10 dilution of household bleach) are effective depending on the amount of organic material (e.g., blood, mucus) present on the surface to be cleaned and disinfected. Commercially available chemical germicides may be more compatible with certain medical devices that might be corroded by repeated exposure to sodium hypochlorite, especially to the 1:10 dilution.

Survival of HIV in the Environment

The most extensive study on the survival of HIV after drying involved greatly concentrated HIV samples, i.e., 10 million tissue-culture infectious doses per milliliter (31). This concentration is at least 100,000 times greater than that typically found in the blood or serum of patients with HIV infection. HIV was detectable by tissue-culture techniques 1-3 days after drying, but the rate of inactivation was rapid. Studies performed at CDC have also shown that drying HIV causes a rapid (within several hours) 1-2 log (90%-99%) reduction in HIV concentration. In tissue-culture fluid, cell-free HIV could be detected for up to 15 days at room temperature, up to 11 days at 37°C (98.6°F), and up to 1 day if the HIV was cell-associated.

When considered in the context of environmental conditions in health-care facilities, these results do not require any changes in currently recommended sterilization, disinfection, or housekeeping strategies. When medical devices are contaminated with blood or other body fluids, existing recommendations include the cleaning of these instruments, followed by disinfection or sterilization, depending on the type of medical device. These protocols assume "worst-case" conditions of extreme virologic and microbiologic contamination, and whether viruses have been inactivated after drying plays no role in formulating these strategies. Consequently, no changes in published procedures for cleaning, disinfecting, or sterilizing need to be made.

Housekeeping

Environmental surfaces such as walls, floors, and other surfaces are not associated with transmission of infections to patients or health-care workers. Therefore, extraordinary attempts to disinfect or sterilize these environmental surfaces are not necessary. However, cleaning and removal of soil should be done routinely.

Cleaning schedules and methods vary according to the area of the hospital or institution, type of surface to be cleaned, and the amount and type of soil present. Horizontal surfaces (e.g., bedside tables and hard-surfaced flooring) in patient-care areas are usually cleaned on a regular basis, when soiling or spills occur, and when a patient is discharged. Cleaning of walls, blinds, and curtains is recommended only if they are visibly soiled. Disinfectant fogging is an unsatisfactory method of decontaminating air and surfaces and is not recommended.

Disinfectant-detergent formulations registered by EPA can be used for cleaning environmental surfaces, but the actual physical removal of microorganisms by scrubbing is probably at least as important as any antimicrobial effect of the cleaning agent used. Therefore, cost, safety, and acceptability by housekeepers can be the main criteria for selecting any such registered agent. The manufacturers' instructions for appropriate use should be followed.

Cleaning and Decontaminating Spills of Blood or Other Body Fluids

Chemical germicides that are approved for use as "hospital disinfectants" and are tuberculocidal when used at recommended dilutions can be used to decontaminate spills of blood and other body fluids. Strategies for decontaminating spills of blood and other body fluids in a patient-care setting are different from those for spills of cultures or other materials in clinical, public health, or research laboratories. In patient-care areas, visible material should first be removed and then the area should be decontaminated. With large spills of cultured or concentrated infectious agents in the laboratory, the contaminated area should be flooded with a liquid germicide before cleaning, then decontaminated with fresh germicidal chemical. In both settings, gloves should be worn during the cleaning and decontaminating procedures.

Laundry

Although soiled linen has been identified as a source of large numbers of certain pathogenic microorganisms, the risk of actual disease transmission is negligible. Rather than rigid procedures and specifications, hygienic and common-sense storage and processing of clean and soiled linen are recommended (26). Soiled linen should be handled as little as possible and with minimum agitation to prevent gross microbial contamination of the air and of persons handling the linen. All soiled linen should be bagged at the location where it was used; it should not be sorted or rinsed in patient-care areas. Linen soiled with blood or body fluids should be placed and transported in bags that prevent leakage. If hot water is used, linen should be washed with detergent in water at least 71°C (160°F) for 25 minutes. If low-temperature (≤70°C (158°F) laundry cycles are used, chemicals suitable for low-temperature washing at proper use concentration should be used.

Infective Waste

There is no epidemiologic evidence to suggest that most hospital waste is any more infective than residential waste. Moreover, there is no epidemiologic evidence that hospital waste has caused disease in the community as a result of improper disposal. Therefore, identifying wastes for which special precautions are indicated is largely a matter of judgment about the relative risk of disease transmission. The most practical approach to the management of infective waste is to identify those wastes with the potential for causing infection during handling and disposal and for which some special precautions appear prudent. Hospital wastes for which special precautions appear prudent include microbiology laboratory waste, pathology waste, and blood specimens or blood products.

While any item that has had contact with blood, exudates, or secretions may be potentially infective, it is not usually considered practical or necessary to treat all such waste as infective (23,26). Infective waste, in general, should either be incinerated or should be autoclaved before disposal in a sanitary landfill. Bulk blood, suctioned fluids, excretions, and secretions may be carefully poured down a drain connected to a sanitary sewer. Sanitary sewers may also be used to dispose of other infectious wastes capable of being ground and flushed into the sewer.

IMPLEMENTATION OF RECOMMENDED PRECAUTIONS

Employers of health-care workers should ensure that policies exist for:

1. Initial orientation and continuing education and training of all health-care workers—including students and trainees—on the epidemiology, modes of transmission, and prevention of HIV and other blood-borne infections and the need for routine use of universal blood and body-fluid precautions for all patients.

2. Provision of equipment and supplies necessary to minimize the risk of infection with HIV and other blood-borne pathogens.

3. Monitoring adherence to recommended protective measures. When monitoring reveals a failure to follow recommended precautions, counseling, education, and/or re-training should be provided, and, if necessary, appropriate disciplinary action should be considered.

Professional associations and labor organizations, through continuing education efforts, should emphasize the need for health-care workers to follow recommended precautions.

SEROLOGIC TESTING FOR HIV INFECTION

Background

A person is identified as infected with HIV when a sequence of tests, starting with repeated enzyme immunoassays (EIAs) and including a Western blot or similar, more specific assay, are repeatedly reactive. Persons infected with HIV usually develop antibody against the virus within 6-12 weeks after infection.

The sensitivity of the currently licensed EIA tests is at least 99% when they are performed under optimal laboratory conditions on serum specimens from persons infected for ≥12 weeks. Optimal laboratory conditions include the use of reliable reagents, provision of continuing education of personnel, quality control of procedures, and participation in performance-evaluation programs. Given this performance, the probability of a false-negative test is remote except during the first several weeks after infection, before detectable antibody is present. The proportion of infected persons with a false-negative test attributed to absence of antibody in the early stages of infection is dependent on both the incidence and prevalence of HIV infection in a population (Table C.1).

The specificity of the currently licensed EIA tests is approximately 99% when repeatedly reactive tests are considered. Repeat testing of initially reactive specimens by EIA is required to reduce the likelihood of laboratory error. To increase further the specificity of serologic tests, laboratories must use a supplemental test, most often the Western blot, to validate repeatedly reactive EIA results. Under optimal laboratory conditions, the sensitivity of the Western blot test is comparable to or greater than that of a repeatedly reactive EIA, and the Western blot is highly specific when strict criteria are used to interpret the test results.

The testing sequence of a repeatedly reactive EIA and a positive Western blot test is highly predictive of HIV infection, even in a population with a low prevalence of infection (Table C.2). If the Western blot test result is indeterminant, the testing sequence is considered equivocal for HIV infection. When this occurs, the Western blot test should be repeated on the same serum sample, and, if still indeterminant, the testing sequence should be repeated on a sample collected 3-6 months later. Use of other supplemental tests may aid in interpreting of results on samples that are persistently indeterminant by Western blot.

TABLE C.1 Estimated Annual Number of Patients Infected with HIV not Detected by HIV-Antibody Testing in a Hypothetical Hospital with 10,000 Admissions/Year*

Beginning prevalence of HIV infection	Annual incidence of HIV infection	Approximate number of HIV-infected patients	Approximate number of HIV-infected patients not detected
5.0%	1.0%	550	17-18
5.0%	0.5%	525	11-12
1.0%	0.2%	110	3-4
1.0%	0.1%	105	2-3
0.1%	0.02%	11	0-1
0.1%	0.01%	11	0-1

*The estimates are based on the following assumptions: 1) the sensitivity of the screening test is 99% (i.e., 99% of HIV-infected persons with antibody will be detected); 2) persons infected with HIV will not develop detectable antibody (seroconvert) until 6 weeks (1.5 months) after infection; 3) new infections occur at an equal rate throughout the year; 4) calculations of the number of HIV-infected persons in the patient population are based on the mid-year prevalence, which is the beginning prevalence plus half the annual incidence of infections.

Testing of Patients

Previous CDC recommendations have emphasized the value of HIV serologic testing of patients for:

1) management of parenteral or mucous-membrane exposures of health-care workers,
2) patient diagnosis and management, and
3) counseling and serologic testing to prevent and control HIV transmission in the community.

In addition, more recent recommendations have stated that hospitals, in conjunction with state and local health departments, should periodically determine the prevalence of HIV infection among patients from age groups at highest risk of infection (32).

Adherence to universal blood and body-fluid precautions recommended for the care of all patients will minimize the risk of transmission of HIV and other blood-borne pathogens from patients to health-care workers. The utility of routine HIV serologic testing of patients as an adjunct to universal precautions is unknown. Results of such testing may not be available in emergency or outpatient settings. In addition, some recently infected patients will not have detectable antibody to HIV (Table C.1).

TABLE C.2 Predictive Value of Positive HIV-Antibody Tests in Hypothetical Populations with Different Prevalences of Infection

	Prevalence of infection	Predictive value of positive test*
Repeatedly reactive enzyme immunoassay (EIA)†	0.2%	28.41%
	2.0%	80.16%
	20.0%	98.02%
Repeatedly reactive EIA followed by positive Western Blot (WB)§	0.2%	99.75%
	2.0%	99.97%
	20.0%	99.99%

*Proportion of persons with positive tests who are actually infected with HIV.
†Assumes EIA sensitivity of 99.0% and specificity of 99.5%.
§Assumes WB sensitivity of 99.0% and specificity of 99.9%.

Personnel in some hospitals have advocated serologic testing of patients in settings in which exposure of health-care workers to large amounts of patients' blood may be anticipated. Specific patients for whom serologic testing has been advocated include those undergoing major operative procedures and those undergoing treatment in critical-care units, especially if they have conditions involving uncontrolled bleeding. Decisions regarding the need to establish testing programs for patients should be made by physicians or individual institutions. In addition, when deemed appropriate, testing of individual patients may be performed on agreement between the patient and the physician providing care.

In addition to the universal precautions recommended for all patients, certain additional precautions for the care of HIV-infected patients undergoing major surgical operations have been proposed by personnel in some hospitals. For example, surgical procedures on an HIV-infected patient might be altered so that hand-to-hand passing of sharp instruments would be eliminated; stapling instruments rather than hand-suturing equipment might be used to perform tissue approximation; electrocautery devices rather than scalpels might be used as cutting instruments; and, even though uncomfortable, gowns that totally prevent seepage of blood onto the skin of members of the operative team might be worn. While such modifications might further minimize the risk of HIV infection for members of the operative team, some of these techniques could result in prolongation of operative time and could potentially have an adverse effect on the patient.

Testing programs, if developed, should include the following principles:

- Obtaining consent for testing.
- Informing patients of test results, and providing counseling for seropositive patients by properly trained persons.
- Assuring that confidentiality safeguards are in place to limit knowledge of test results to those directly involved in the care of infected patients or as required by law.
- Assuring that identification of infected patients will not result in denial of needed care or provision of suboptimal care.
- Evaluating prospectively: 1) the efficacy of the program in reducing the incidence of parenteral, mucous-membrane, or significant cutaneous exposures of health-care workers to the blood or other body fluids of HIV-infected patients and 2) the effect of modified procedures on patients.

Testing of Health-Care Workers

Although transmission of HIV from infected health-care workers to patients has not been reported, transmission during invasive procedures remains a possibility. Transmission of hepatitis B virus (HBV)—a blood-borne agent with a considerably greater potential for nosocomial spread—from health-care workers to patients has been documented. Such transmission has occurred in situations (e.g., oral and gynecologic surgery) in which health-care workers, when tested, had very high concentrations of HBV in their blood (at least 100 million infectious virus particles per milliliter, a concentration much higher than occurs with HIV infection), and the health-care workers sustained a puncture wound while performing invasive procedures or had exudative or weeping lesions or microlacerations that allowed virus to contaminate instruments or open wounds of patients (33,34).

The hepatitis B experience indicates that only those health-care workers who perform certain types of invasive procedures have transmitted HBV to patients. Adherence to recommendations in this document will minimize the risk of transmission of HIV and other blood-borne pathogens from health-care workers to patients during invasive procedures. Since transmission of HIV from infected health-care workers performing invasive procedures to their patients has not been reported and would be expected to occur only very rarely, if at all, the utility of routine testing of such health-care workers to prevent transmission of HIV cannot be assessed. If consideration is given to developing a serologic testing program for health-care workers who perform invasive procedures, the frequency of testing, as well as the issues of consent and confidentiality, and consequences of test results—as previously outlined for testing programs for patients—must be addressed.

MANAGEMENT OF INFECTED HEALTH-CARE WORKERS

Health-care workers with impaired immune systems resulting from HIV infection or other causes are at increased risk of acquiring or experiencing serious complications of infectious disease. Of particular concern is the risk of severe infection following exposure to patients with infectious diseases that are easily transmitted if appropriate precautions are not taken (e.g., measles, varicella). Any health-care worker with an impaired immune system should be counseled about the potential risk associated with taking care of patients with any transmissible infection and should continue to follow existing recommendations for infection control to minimize risk of exposure to other infectious agents (7,35). Recommendations of the Immunization Practices Advisory Committee (ACIP) and institutional policies concerning requirements for vaccinating health-care workers with live-virus vaccines (e.g., measles, rubella) should also be considered.

The question of whether workers infected with HIV—especially those who perform invasive procedures—can adequately and safely be allowed to perform patient-care duties or whether their work assignments should be changed must be determined on an individual basis. These decisions should be made by the health-care worker's personal physician(s) in conjunction with the medical directors and personnel health service staff of the employing institution or hospital.

MANAGEMENT OF EXPOSURES

If a health-care worker has a parenteral (e.g., needlestick or cut) or mucous-membrane (e.g., splash to the eye or mouth) exposure to blood or other body fluids or has a cutaneous exposure involving large amounts of blood or prolonged contact with blood—especially when the exposed skin is chapped, abraded, or afflicted with dermatitis—the source patient should be informed of the incident and tested for serologic evidence of HIV infection after consent is obtained. Policies should be developed for testing source patients in situations in which consent cannot be obtained (e.g., an unconscious patient).

If the source patient has AIDS, is positive for HIV antibody, or refuses the test, the health-care worker should be counseled regarding the risk of infection and evaluated clinically and serologically for evidence of HIV infection as soon as possible after the exposure. The health-care worker should be advised to report and seek medical evaluation for any acute febrile illness that occurs within 12 weeks after the exposure. Such an illness—particularly one characterized by fever, rash, or lymphadenopathy—may be indicative of recent HIV infection. Seronegative health-care workers should be retested 6 weeks post-exposure and on a periodic basis thereafter (e.g., 12 weeks and 6 months after exposure) to determine whether transmission has occurred. During this follow-up period—especially the first 6-12 weeks after exposure, when most infected persons are expected to seroconvert—exposed health-care workers should follow U.S. Public Health Service (PHS) recommendations for preventing transmission of HIV (36,37).

No further follow-up of a health-care worker exposed to infection as described above is necessary if the source patient is seronegative unless the source patient is at high risk of HIV infection. In the latter case, a subsequent specimen (e.g., 12 weeks following exposure) may be obtained from the health-care worker for antibody testing. If the source patient cannot be identified, decisions regarding appropriate follow-up should be individualized. Serologic testing should be available to all health-care workers who are concerned that they may have been infected with HIV.

If a patient has a parenteral or mucous-membrane exposure to blood or other body fluid of a health-care worker, the patient should be informed of the incident, and the same procedure outlined above for management of exposures should be followed for both the source health-care worker and the exposed patient.

REFERENCES

1. CDC. Acquired immunodeficiency syndrome (AIDS): Precautions for clinical and laboratory staffs. MMWR 1982;31:577-80.

2. CDC. Acquired immunodeficiency syndrome (AIDS): Precautions for health-care workers and allied professionals. MMWR 1983;32:450-1.

3. CDC. Recommendations for preventing transmission of infection with human T-lymphotropic virus type III/lymphadenopathy-associated virus in the workplace. MMWR 1985;34:681-6,691-5.

4. CDC. Recommendations for preventing transmission of infection with human T-lym-

photropic virus type III/lymphadenopathy-associated virus during invasive procedures. MMWR 1986;35:221-3.

5. CDC. Recommendations for preventing possible transmission of human T-lymphotropic virus type III/lymphadenopathy-associated virus from tears. MMWR 1985;34:533-4.

6. CDC. Recommendations for providing dialysis treatment to patients infected with human T-lymphotropic virus type III/lymphadenopathy-associated virus infection. MMWR 1986;35:376-8, 383.

7. Garner JS, Simmons BP. Guideline for isolation precautions in hospitals. Infect Control 1983;4(suppl):245-325.

8. CDC. Recommended infection control practices for dentistry. MMWR 1986;35:237-42.

9. McCray E, The Cooperative Needlestick Surveillance Group. Occupational risk of the acquired immunodeficiency syndrome among health care workers. N Engl J Med 1986;314:1127-32.

10. Henderson DK, Saah AJ, Zak BJ, et al. Risk of nosocomial infection with human T-cell lymphotropic virus type III/lymphadenopathy-associated virus in a large cohort of intensively exposed health care workers. Ann Intern Med 1986;104:644-7.

11. Gerberding JL, Bryant-LeBlanc CE, Nelson K, et al. Risk of transmitting the human immunodeficiency virus, cytomegalovirus, and hepatitis B virus to health care workers exposed to patients with AIDS and AIDS-related conditions. J Infect Dis 1987;156:1-8.

12. McEvoy M, Porter K, Mortimer P, Simmons N, Shanson D. Prospective study of clinical, laboratory, and ancillary staff with accidental exposures to blood or other body fluids from patients infected with HIV. Br Med J 1987;294:1595-7.

13. Anonymous. Needlestick transmission of HTLV-III from a patient infected in Africa. Lancet 1984;2:1376-7.

14. Oksenhendler E, Harzic M, Le Roux JM, Rabian C, Clauvel JP. HIV infection with seroconversion after a superficial needlestick injury to the finger. N Engl J Med 1986;315:582.

15. Neisson-Vernant C, Arfi S, Mathez D, Leibowitch J, Monplaisir N. Needlestick HIV seroconversion in a nurse. Lancet 1986;2:814.

16. Grint P, McEvoy M. Two associated cases of the acquired immune deficiency syndrome (AIDS). PHLS Commun Dis Rep 1985;42:4.

17. CDC. Apparent transmission of human T-lymphotropic virus type III/lymphadenopathy-associated virus from a child to a mother providing health care. MMWR 1986;35:76-9.

18. CDC. Update: Human immunodeficiency virus infections in health-care workers exposed to blood of infected patients. MMWR 1987;36:285-9.

19. Kline RS, Phelan J, Friedland GH, et al. Low occupational risk for HIV infection for dental professionals (Abstract). In: Abstracts from the III International Conference on AIDS, 1-5 June 1985. Washington, DC: 155.

20. Baker JL, Kelen GD, Sivertson KT, Quinn TC. Unsuspected human immunodeficiency virus in critically ill emergency patients. JAMA 1987;257:2609-11.

21. Favero MS. Dialysis-associated diseases and their control. In: Bennett JV, Brachman PS, eds. Hospital infections. Boston: Little, Brown and Company, 1985:267-84.

22. Richardson JH, Barkley WE, eds. Biosafety in microbiological and biomedical laboratories, 1984. Washington, DC: US Department of Health and Human Services, Public Health Service. HHS publication no. (CDC) 84-8395.

23. CDC. Human T-lymphotropic virus type III/ lymphadenopathy-associated virus: Agent summary statement. MMWR 1986;35:540-2,547-9.

24. Environmental Protection Agency. EPA guide for infectious waste management. Washington, DC: US Environmental Protection Agency, May 1986 (Publication no. EPA/530-SW-86-014).

25. Favero MS. Sterilization, disinfection, and antisepsis in the hospital. In: Manual of clinical microbiology. 4th ed. Washington, DC: American Society for Microbiology, 1985;129-37.

26. Garner JS, Favero MS. Guideline for handwashing and hospital environmental control, 1985. Atlanta: Public Health Service, Centers for Disease Control, 1985. HHS publication no. 99-1117.

27. Spire B, Montagnier L, Barre-Sinoussi F, Chermann JC. Inactivation of lymphadenopathy associated virus by chemical disinfectants. Lancet 1984;2:899-901.

28. Martin LS, McDougal JS, Loskoski SL. Disinfection and inactivation of the human T lymphotropic virus type III/lymphadenopathy-associated virus. J Infect Dis 1985;152:400-3.

29. McDougal JS, Martin LS, Cort SP, et al. Thermal inactivation of the acquired immunodeficiency syndrome virus-III/lymphadenopathy-associated virus, with special reference to antihemophilic factor. J Clin Invest 1985;76:875-7.

30. Spire B, Barre-Sinoussi F, Dormont D, Montagnier L, Chermann JC. Inactivation of lymphadenopathy-associated virus by heat, gamma rays, and ultraviolet light. Lancet 1985;1:188-9.

31. Resnik L, Veren K, Salahuddin SZ, Tondreau S, Markham PD. Stability and inactivation of HTLV-III/LAV under clinical and laboratory environments. JAMA 1986;255:1887-91.

32. CDC. Public Health Service (PHS) guidelines for counseling and antibody testing to prevent HIV infection and AIDS. MMWR 1987;3:509-15.

33. Kane MA, Lettau LA. Transmission of HBV from dental personnel to patients. J Am Dent Assoc 1985;110:634-6.

34. Lettau LA, Smith JD, Williams D, et. al. Transmission of hepatitis B with resultant restriction of surgical practice. JAMA 1986;255:934-7.

35. Williams WW. Guideline for infection control in hospital personnel. Infect Control 1983; 4(suppl):326-49.

36. CDC. Prevention of acquired immune deficiency syndrome (AIDS): Report of interagency recommendations. MMWR 1983; 32:101-3.

37. CDC. Provisional Public Health Service interagency recommendations for screening donated blood and plasma for antibody to the virus causing acquired immunodeficiency syndrome. MMWR 1985;34:1-5.

II. UPDATE: UNIVERSAL PRECAUTIONS FOR PREVENTION OF TRANSMISSION OF HUMAN IMMUNODEFICIENCY VIRUS, HEPATITIS B VIRUS, AND OTHER BLOODBORNE PATHOGENS IN HEALTH-CARE SETTINGS

INTRODUCTION

The purpose of this report is to clarify and supplement the CDC publication entitled "Recommendations for Prevention of HIV Transmission in Health-Care Settings" (1).*****

*****The August 1987 publication (see pages 155-168 of this report) should be consulted for general information and specific recommendations not addressed in this update.

In 1983, CDC published a document entitled "Guideline for Isolation Precautions in Hospitals" (2) that contained a section entitled "Blood and Body Fluid Precautions." The recommendations in this section called for blood and body fluid precautions when a patient was known or suspected to be infected with bloodborne pathogens. In August 1987, CDC published a document entitled "Recommendations for Prevention of HIV Transmission in Health-Care Settings" (1). In contrast to the 1983 document, the 1987 document recommended that blood and body fluid precautions be consistently used for all patients regardless of their bloodborne infection status. This extension of blood and body fluid precautions to *all* patients is referred to as "Universal Blood and Body Fluid Precautions" or "Universal Precautions." Under universal precautions, blood and certain body fluids of all patients are considered potentially infectious for human immunodeficiency virus (HIV), hepatitis B virus (HBV), and other bloodborne pathogens.

Universal precautions are intended to prevent parenteral, mucous membrane, and nonintact skin exposures of health-care workers to bloodborne pathogens. In addition, immunization with HBV vaccine is recommended as an important adjunct to universal precautions for health-care workers who have exposures to blood (3,4).

Since the recommendations for universal precautions were published in August 1987, CDC and the Food and Drug Administration (FDA) have received requests for clarification of the following issues: 1) body fluids to which universal precautions apply, 2) use of protective barriers, 3) use of gloves for phlebotomy, 4) selection of gloves for use while observing universal precautions, and 5) need for making changes in waste management programs as a result of adopting universal precautions.

BODY FLUIDS TO WHICH UNIVERSAL PRECAUTIONS APPLY

Universal precautions apply to blood and to other body fluids containing visible blood. Occupational transmission of HIV and HBV to health-care workers by blood is documented (4,5). *Blood is the single most important source of HIV, HBV, and other bloodborne pathogens in the occupational setting. Infection control efforts for HIV, HBV, and other bloodborne pathogens must focus on preventing exposures to blood as well as on delivery of HBV immunization.*

Universal precautions also apply to semen and vaginal secretions. Although both of these fluids have been implicated in the sexual transmission of HIV and HBV, they have not been implicated in occupational transmission from patient to health-care worker. This observation is not unexpected, since exposure to semen in the usual health-care setting is limited, and the routine practice of wearing gloves for performing vaginal examinations protects health-care workers from exposure to potentially infectious vaginal secretions.

Universal precautions also apply to tissues and to the following fluids: cerebrospinal fluid (CSF), synovial fluid, pleural fluid, peritoneal fluid, pericardial fluid, and amniotic fluid. The risk of transmission of HIV and HBV from these fluids is unknown; epidemiologic studies in the health-care and community setting are currently inadequate to assess the potential risk to health-care workers from occupational exposures to them. However, HIV has been isolated from CSF, synovial, and amniotic fluid (6-8), and HBsAg has been detected in synovial fluid, amniotic fluid, and peritoneal fluid (9-11). One case of HIV transmission was reported after a percutaneous exposure to bloody pleural fluid obtained by needle aspiration (12). Whereas aseptic procedures used to obtain these fluids for diagnostic or therapeutic purposes protect health-care workers from skin exposures, they cannot prevent penetrating injuries due to contaminated needles or other sharp instruments.

BODY FLUIDS TO WHICH UNIVERSAL PRECAUTIONS DO NOT APPLY

Universal precautions do not apply to feces, nasal secretions, sputum, sweat, tears, urine, and vomitus unless they contain visible blood. The risk of transmission of HIV and HBV from these fluids and materials is extremely low or nonexistent. HIV has been isolated and HBsAg has been demonstrated in some of these fluids; however, epidemiologic studies in the health-care and community setting have not implicated these fluids or materials in the transmission of HIV and HBV infections (13,14). Some of the above fluids and excretions represent a potential source for nosocomial and community-acquired infections with other pathogens, and recommendations for preventing the transmission of non-bloodborne pathogens have been published (2).

PRECAUTIONS FOR OTHER BODY FLUIDS IN SPECIAL SETTINGS

Human breast milk has been implicated in perinatal transmission of HIV, and HBsAg has been found in the milk of mothers infected with HBV (10,13). However, occupational exposure to human breast milk has not been implicated in the transmission of HIV or HBV infection to health-care workers. Moreover, the health-care worker will not have the same type of intensive exposure to breast milk as the nursing neonate. Whereas universal precautions do not apply to human breast milk, gloves may be worn by health-care workers in situations where exposures to breast milk might be frequent, for example, in breast milk banking.

Saliva of some persons infected with HBV has been shown to contain HBV-DNA at concentrations 1/1,000 to 1/10,000 of that found in the infected person's serum (15). HBsAg-positive saliva has been shown to be infectious when injected into experimental animals and in human bite exposures (16-18). However, HBsAg-positive saliva has not been shown to be infectious when applied to oral mucous membranes in experimental primate studies (18) or through contamination of musical instruments or cardiopulmonary resuscitation dummies used by HBV carriers (19,20).

Epidemiologic studies of nonsexual household contacts of HIV-infected patients, including several small series in which HIV transmission failed to occur after bites or after percutaneous inoculation or contamination of cuts and open wounds with saliva from HIV-infected patients, suggest that the potential for salivary transmission of HIV is remote (5,13,14,21,22). One case report from Germany has suggested the possibility of transmission of HIV in a household setting from an infected child to a sibling through a human bite (23). The bite did not break the skin or result in bleeding. Since the date of seroconversion to HIV was not known for either child in this case, evidence for the role of saliva in the transmission of virus is unclear (23). Another case report suggested the possibility of transmission of HIV from husband to wife by contact with saliva during kissing (24). However, follow-up studies did not confirm HIV infection in the wife (21).

Universal precautions do not apply to saliva. General infection control practices already in existence—including the use of gloves for digital examination of mucous membranes and endotracheal suctioning, and handwashing after exposure to saliva—should further minimize the minute risk, if any, for salivary transmission of HIV and HBV (1,25). Gloves need not be worn when feeding patients and when wiping saliva from skin.

Special precautions, however, are recommended for dentistry (1). Occupationally acquired infection with HBV in dental workers has been documented (4), and two possible cases of occupationally acquired HIV infection involving dentists have been reported (5,26). During dental procedures, contamination of saliva with blood is predictable, trauma to health-care workers' hands is common, and blood spattering may occur. Infection control precautions for dentistry minimize the potential for non-intact skin and mucous membrane contact of dental health-care workers to blood-contaminated saliva of patients. In addition, the use of gloves for oral examinations and treatment in the dental setting may also protect the patient's oral mucous membranes from exposures to blood, which may occur from breaks in the skin of dental workers' hands.

USE OF PROTECTIVE BARRIERS

Protective barriers reduce the risk of exposure of the health-care worker's skin or mucous membranes to potentially infective materials. For universal precautions, protective barriers reduce the risk of exposure to blood, body fluids containing visible blood, and other fluids to which universal precautions apply.

Examples of protective barriers include gloves, gowns, masks, and protective eyewear. Gloves should reduce the incidence of contamination of hands, but they cannot prevent penetrating injuries due to needles or other sharp instruments. Masks and protective eyewear or face shields should reduce the incidence of contamination of mucous membranes of the mouth, nose, and eyes.

Universal precautions are intended to supplement rather than replace recommendations for routine infection control, such as handwashing and using gloves to prevent gross microbial contamination of hands (27). Because specifying the types of barriers needed for every possible clinical situation is impractical, some judgment must be exercised.

The risk of nosocomial transmission of HIV, HBV, and other bloodborne pathogens can be minimized if health-care workers use the following general guidelines:******

1. Take care to prevent injuries when using needles, scalpels, and other sharp instruments or devices; when handling sharp instruments after procedures; when cleaning used instruments; and when disposing of used needles. Do not recap used needles by hand; do not remove used needles from disposable syringes by hand; and do not bend, break, or otherwise manipulate used needles by hand. Place used disposable syringes and needles, scalpel blades, and other sharp items in puncture-resistant containers for disposal. Locate the puncture-resistant containers as close to the use area as is practical.

2. Use protective barriers to prevent exposure to blood, body fluids containing visible blood, and other fluids to which universal precautions apply. The type of protective barrier(s) should be appropriate for the procedure being performed and the type of exposure anticipated.

3. Immediately and thoroughly wash hands and other skin surfaces that are contaminated with blood, body fluids containing visible blood, or other body fluids to which universal precautions apply.

GLOVE USE FOR PHLEBOTOMY

Gloves should reduce the incidence of blood contamination of hands during phlebotomy (drawing blood samples), but they cannot prevent penetrating injuries caused by needles or other sharp instruments. The likelihood of hand contamination with blood containing HIV, HBV, or other bloodborne pathogens during phlebotomy depends on several factors:

1) the skill and technique of the health-care worker;

2) the frequency with which the health-care worker performs the procedure (other factors being equal, the cumulative risk of blood exposure is higher for a health-care worker who performs more procedures);

3) whether the procedure occurs in a routine or emergency situation (where blood contact may be more likely); and,

4) the prevalence of infection with bloodborne pathogens in the patient population.

The likelihood of infection after skin exposure to blood containing HIV or HBV will depend on the concentration of virus (viral concentration is much higher for hepatitis B than for HIV), the duration of contact, the presence of skin lesions on the hands of the health-care worker, and—for HBV—the immune status of the health-care worker. Although not accurately quantified, the risk of HIV infection following intact skin contact with infective blood is certainly much less than the 0.5% risk following percutaneous needle stick exposures (5). In universal precautions, all blood is assumed to be potentially infective for bloodborne pathogens, but in certain settings (e.g., volunteer blood-donation centers) the prevalence of infection with some bloodborne pathogens (e.g., HIV, HBV) is known to be very low. Some institutions have relaxed recommendations for using gloves for phlebotomy procedures by skilled phlebotomists in settings where the prevalence of bloodborne pathogens is known to be very low.

Institutions that judge that routine gloving for all phlebotomies is not necessary should periodically reevaluate their policy. Gloves should always be available to health-care workers who wish to use them for phlebotomy. In addition, the following general guidelines apply:

1. Use gloves for performing phlebotomy when the health-care worker has cuts, scratches, or other breaks in his/her skin.

2. Use gloves in situations where the health-care worker judges that hand contamination with blood may occur, for example, when performing phlebotomy on an uncooperative patient.

3. Use gloves for performing finger and/or heel sticks on infants and children.

4. Use gloves when persons are receiving training in phlebotomy.

SELECTION OF GLOVES

The Center for Devices and Radiological Health, FDA, has responsibility for regulating the medical

******The August 1987 publication (see pages 155-168 of this report) should be consulted for general information and specific recommendations not addressed in this update.

glove industry. Medical gloves include those marketed as sterile surgical or nonsterile examination gloves made of vinyl or latex. General purpose utility ("rubber") gloves are also used in the health-care setting, but they are not regulated by FDA since they are not promoted for medical use. There are no reported differences in barrier effectiveness between intact latex and intact vinyl used to manufacture gloves. Thus, the type of gloves selected should be appropriate for the task being performed.

The following general guidelines are recommended:

1. Use sterile gloves for procedures involving contact with normally sterile areas of the body.
2. Use examination gloves for procedures involving contact with mucous membranes, unless otherwise indicated, and for other patient care or diagnostic procedures that do not require the use of sterile gloves.
3. Change gloves between patient contacts.
4. Do not wash or disinfect surgical or examination gloves for reuse. Washing with surfactants may cause "wicking," i.e., the enhanced penetration of liquids through undetected holes in the glove. Disinfecting agents may cause deterioration.
5. Use general-purpose utility gloves (e.g., rubber household gloves) for housekeeping chores involving potential blood contact and for instrument cleaning and decontamination procedures. Utility gloves may be decontaminated and reused but should be discarded if they are peeling, cracked, or discolored, or if they have punctures, tears, or other evidence of deterioration.

WASTE MANAGEMENT

Universal precautions are not intended to change waste management programs previously recommended by CDC for health-care settings (1). Policies for defining, collecting, storing, decontaminating, and disposing of infective waste are generally determined by institutions in accordance with state and local regulations. Information regarding waste management regulations in health-care settings may be obtained from state or local health departments or agencies responsible for waste management.

Reported by:

Center for Devices and Radiological Health, Food and Drug Administration.

Hospital Infections Program, AIDS Program, and Hepatitis Branch, Division of Viral Diseases, Center for Infectious Diseases, National Institute for Occupational Safety and Health, CDC.

MMWR EDITORIAL NOTE

Implementation of universal precautions does not eliminate the need for other category- or disease-specific isolation precautions, such as enteric precautions for infectious diarrhea or isolation for pulmonary tuberculosis (1,2). In addition to universal precautions, detailed precautions have been developed for the following procedures and/or settings in which prolonged or intensive exposures to blood occur: invasive procedures, dentistry, autopsies or morticians' services, dialysis, and the clinical laboratory. These detailed precautions are found in the August 21, 1987, "Recommendations for Prevention of HIV Transmission in Health-Care Settings" (1). In addition, specific precautions have been developed for research laboratories (28).

REFERENCES

1. Centers for Disease Control. Recommendations for prevention of HIV transmission in health-care settings. MMWR 1987;36(suppl no. 2S),3S-18S.

2. Garner JS, Simmons BP. Guideline for isolation precautions in hospitals. Infect Control 1983:4;245-325.

3. Immunization Practices Advisory Committee. Recommendations for protection against viral hepatitis. MMWR 1985;34:313-24,329-35.

4. Department of Labor, Department of Health and Human Services. Joint advisory notice: protection against occupational exposure to hepatitis B virus (HBV) and human immunodeficiency virus (HIV). Washington, DC:US

Department of Labor, US Department of Health and Human Services, 1987.

5. Centers for Disease Control. Update: Acquired immunodeficiency syndrome and human immunodeficiency virus infection among health-care workers. MMWR 1988;37:229-34,239.

6. Hollander H, Levy JA. Neurologic abnormalities and recovery of human immunodeficiency virus from cerebrospinal fluid. Ann Intern Med 1987;106:692-5.

7. Wirthrington RH, Cornes P, Harris JRW, et al. Isolation of human immunodeficiency virus from synovial fluid of a patient with reactive arthritis. Br Med J 1987;294:484.

8. Mundy DC, Schinazi RF, Gerber AR, Nahmias AJ, Randall HW. Human immunodeficiency virus isolated from amniotic fluid. Lancet 1987;2:459-60.

9. Onion DK, Crumpacker CS, Gilliland BC. Arthritis of hepatitis associated with Australia antigen. Ann Intern Med 1971;75:29-33.

10. Lee AKY, Ip HMH, Wong VCW. Mechanisms of maternal-fetal transmission of hepatitis B virus. J Infect Dis 1978;138:668-71.

11. Bond WW, Petersen NJ, Gravelle CR, Favero MS. Hepatitis B virus in peritoneal dialysis fluid: A potential hazard. Dialysis and Transplantation 1982;11:592-600.

12. Oskenhendler E, Harzic M, Le Roux J-M, Rabian C, Clauvel JP. HIV infection with seroconversion after a superficial needlestick injury to the finger (Letter). N Engl J Med 1986;315:582.

13. Lifson AR. Do alternate modes for transmission of human immunodeficiency virus exist? A review. JAMA 1988;259:1353-6.

14. Friedland GH, Saltzman BR, Rogers MF, et al. Lack of transmission of HTLV-III/LAV infection to household contacts of patients with AIDS or AIDS-related complex with oral candidiasis. N Engl J Med 1986;314:344-9.

15. Jenison SA, Lemon SM, Baker LN, Newbold JE. Quantitative analysis of hepatitis B virus DNA in saliva and semen of chronically infected homosexual men. J Infect Dis 1987; 156:299-306.

16. Cancio-Bello TP, de Medina M, Shorey J, Valledor MD, Schiff ER. An institutional outbreak of hepatitis B related to a human biting carrier. J Infect Dis 1982;146:652-6.

17. MacQuarrie MB, Forghani B, Wolochow DA. Hepatitis B transmitted by a human bite. JAMA 1974;230:723-4.

18. Scott RM, Snitbhan R, Bancroft WH, Alter HJ, Tingpalapong M. Experimental transmission of hepatitis B virus by semen and saliva. J Infect Dis 1980;142:67-71.

19. Glaser JB, Nadler JP. Hepatitis B virus in a cardiopulmonary resuscitation training course: Risk of transmission from a surface antigen-positive participant. Arch Intern Med 1985; 145:1653-5.

20. Osterholm MT, Bravo ER, Crosson JT, et al. Lack of transmission of viral hepatitis type B after oral exposure to HBsAg-positive saliva. Br Med J 1979;2:1263-4.

21. Curran JW, Jaffe HW, Hardy AM, et al. Epidemiology of HIV infection and AIDS in the United States. Science 1988;239:610-6.

22. Jason JM, McDougal JS, Dixon G, et al. HTLV-III/LAV antibody and immune status of household contacts and sexual partners of persons with hemophilia. JAMA 1986;255:212-5.

23. Wahn V, Kramer HH, Voit T, Bruster HT, Scrampical B, Scheid A. Horizontal transmission of HIV infection between two siblings (Letter). Lancet 1986;2:694.

24. Salahuddin SZ, Groopman JE, Markham PD, et al. HTLV-III in symptom-free seronegative persons. Lancet 1984;2:1418-20.

25. Simmons BP, Wong ES. Guideline for prevention of nosocomial pneumonia. Atlanta: US Department of Health and Human Services, Public Health Service, Centers for Disease Control, 1982.

26. Klein RS, Phelan JA, Freeman K, et al. Low occupational risk of human immunodeficiency virus infection among dental professionals. N Engl J Med 1988;318:86-90.

27. Garner JS, Favero MS. Guideline for handwashing and hospital environmental control, 1985. Atlanta: US Department of Health and Human Services, Public Health Service, Centers for Disease Control, 1985; HHS publication no. 99-1117.

28. Centers for Disease Control. 1988 Agent summary statement for human immunodeficiency virus and report on laboratory-acquired infection with human immunodeficiency virus. MMWR 1988;37(suppl no. S4):1S-22S.

Appendix D

Summary of Zoonotic Pathogens Causing Disease in Man

Microorganism Group	Disease	Agent	Animal Source[a] A	F	B	D	P	C	E	N	L	M	O	S	W	Route of Infection[b] Ae	Ab	Bs	Co	In	Sp	Wi
Gram negative bacteria	Brucellosis	*Brucella* spp.			X	X		X	X				X	X	X	X			X	X	?	
	Colibacillosis	Arizona grp., *E. coli*, *Shigella* spp.	X	X	X	X			X	X	X	X	X	X	X				X	X		
	Enteritis	*Campylobacter* spp.	X	X	X	X		X		X			X	X					X	X		
	Glanders	*Pseudomonas mallei*		X		X		X	X				X		X	X			X	X		X
	Melioidosis	*Pseudomonas pseudomallei*	X	X	X	X		X	X	X	X	X	X	X	X	X	X		X	X		X
	Non-specific purulent anaerobic infections	*Bacteroides* spp., *Fusobacterium* spp.	X		X	X		X	X		X		X	X				X	X			X
	Pasteurellosis	*Pasteurella* spp.	X	X	X	X		X	X			X	X	X				X	X	X		X
	Plague	*Yersinia pestis*		X		X					X	X		X	X	X	X					X
	Ratbite fever	*Spirillum* spp., *Streptobacillus* spp.		X								X						X		X		
	Salmonellosis	*Salmonella* spp.	X	X	X	X	X	X	X	X	X	X	X	X	X	X			X	X		X
	Tularemia	*Francisella tularensis*		X	X	X		X	X		X	X	X		X	X	X	X	X	X	X	X
	Vibriosis	*Vibrio* spp.			X		X	X					X							X		
	Yersiniosis	*Yersinia* spp.	X	X	X	X		X	X		X	X	X	X		?		?	X	X		
Gram positive bacteria	Anthrax	*Bacillus anthracis*		X	X	X		X	X				X	X		X	X		X	X		X
	Botulism	*Clostridium botulinum*	X				X													X		
	Corynebacterial infections	*Corynebacterium* spp.			X			X	X				X						X	X		X
	Erysipeloid	*Erysipelothrix rhusiopathiae*	X				X					X		X					X			

Gram positive bacteria (continued)	Food poisoning	*Clostridium perfringens*	X	X	X	X	X	X	X	X	X	X	X	X	X						X		X
	Gas gangrene, malignant edema	*Clostridium* spp.	X	X	X	X		X	X	X	X	X	X	X	X								X
	Listeriosis[c]	*Listeria monocytogenes*	X	X	X	X	X	X	X			X	X	X	X					X	X		X
	Staphylococcosis	*Staphylococcus aureus*	X		X									X						X	X		
	Streptococcosis	*Streptococcus* spp.	X		X									X						X	X		
	Tetanus	*Clostridium tetani*			X			X	X				X	X	X								X
Mycobacteria	Leprosy[d]	*Mycobacterium leprae*													X					?			
	Mycobacterial infections	*Mycobacterium* spp.	X		?		X							?			X			X			X
	Tuberculosis	*Mycobacterium* spp.	X	X	X	X		X	X	X			X	X	X		X			X	X		
Actinomycetes	Actinomycosis	*Actinomyces* spp.		X	X	X			X					X									X
	Dermatophilosis	*Dermatophilus congolensis*			X			X	X				X		X		X					X	
	Nocardiosis	*Nocardia* spp.		X	X	X		?	X				?	X						X			X
Rickettsiaceae	Boutonneuse fever	*Rickettsia conori*	?			X					X	X			X			X					
	Ehrlichiosis	*Ehrlichia canis*				X												X					
	Epidemic typhus	*Rickettsia prowazekii*													X			X					
	Murine typhus	*Rickettsia typhi*										X			X			X			X		X
	North Asian tick typhus	*Rickettsia sibirica*				X						X	X		X		?	X					?

Continued

Microorganism Group	Disease	Agent	Animal Source[a]													Route of Infection[b]						
			A	F	B	D	P	C	E	N	L	M	O	S	W	Ae	Ab	Bs	Co	In	Sp	Wi
Rickettsiaceae (continued)	Psittacosis, ornithosis	*Chlamydia psittaci*	X	X	X			X				X	X		X	X			X			
	Q fever	*Coxiella burnetii*	X		X			X				X	X		X	X	X		X	X	X	X
	Queensland tick typhus	*Rickettsia australis*										X			X		X					
	Rickettsialpox	*Rickettsia akari*										X				X	X					
	Rocky Mountain spotted fever	*Rickettsia rickettsiae*	?			X					X	X			X	X	X				X	X
	Scrub typhus	*Rickettsia tsutsugamushi*	X									X					X					
Spirochetes	Endemic relapsing fever	*Borrelia* spp.			X				X	X		X			X		X		?			X
	Leptospirosis[c]	*Leptospira* spp.	?	X	X	X	?	X	X			X	X	X	X			X	X	?	X	X
	Lyme disease	*Borrelia burgdorferi*										X			X		X					
Mycoses	Aspergillosis	*Aspergillus fumigatus*	X	X	X	X		X	X			X	X	X		X						
	Blastomycosis	*Blastomyces dermatiditis*		X		X			X						X	X			X			X
	Candidiasis	*Candida* spp.	X		X	X			X				X						X			X
	Coccidioidomycosis	*Coccidioides immitis*		X	X	X		X	X			X	X	X		X			X			
	Cryptococcosis	*Cryptococcus neoformans*	X	X	X	X		X	X			X	X			X			X			
	Geotrichosis	*Geotrichum candidum*			X					X						X			X			
	Histoplasmosis	*Histoplasma capsulatum*		X	X	X			X			X			X	X			X			
	Piedra	*Piedraia* spp., *Trichosporon* spp.			X				X	X									X			

Mycoses (continued)	Phycomycosis[d]	*Absidia* spp., *Basidiobolus* spp., *Entomophthora* spp., *Mortierella* spp., *Mucor* spp., *Rhizopus* spp.			X	X		X	X			X	X	X			X				X		
	Rhinosporidiosis[d]	*Rhinosporidium seeberi*	X		X	X			X											?			
	Ringworm	*Epidermophyton* spp., *Microsporum* spp., *Trichophyton* spp.	X	X	X	X		X	X	X	X	X	X	X	X					X			
	Rhodotorulosis[d]	*Rhodotourula mucilaginosa*	X																	X			
	Sporotrichosis	*Sporothrix schenckii*		X	X	X			X			X		X						X			X
Trematodes	Amphistomiasis	*Gastrodiscoides* spp., *Watsonius* spp.										X		X							X		
	Clinostomiasis[d]	*Clinostomum complanatum*					X														X		
	Clonorchiasis	*Clonorchis* spp., *Opisthorchis* spp., *Amphimerus* spp., *Metorchis* spp., *Pseudamphistomum* spp.	X	X		X	X							X	X						X		
	Cyathocotyliasis[d]	*Prohemistomum vivax*	X	X		X	X														X		
	Dicrocoeliasis[d]	*Dicrocoelium* spp., *Eurytrema pancreaticum*			X			X					X		X						X		
	Echinostomiasis	*Echinostoma* spp., *Artyfechinostomum* spp., *Echinochasmus* spp., *Echinoparyphium* spp., *Euparyphium* spp., *Himasthla* spp., *Hypoderaeum* spp.	X	X		X	X					X									X		

Continued

Microorganism Group	Disease	Agent	Animal Source[a]													Route of Infection[b]						
			A	F	B	D	P	C	E	N	L	M	O	S	W	Ae	Ab	Bs	Co	In	Sp	Wi
Trematodes (continued)	Fascioliasis, fasciolopsiasis	*Fascioloa* spp., *Fasciolopsis* spp.			X	X		X					X	X						X		
	Heterophyiasis	*Heterophyes* spp., *Apophallus* spp., *Centrocestus* spp., *Haplorchis* spp., *Pigydiopsis* spp., *Procerovum* spp., *Stellantchasmus* spp., *Strictodora* spp.	X	X		X	X								X					X		
	Metagonimiasis	*Metagonimus* spp., *Troglotrema* spp.	X	X		X	X							X	X					X		
	Paragonimiasis	*Paragonimus* spp.		X		X	X			X				X	X					X		
	Schistosomiasis, cercarial dermatitis	*Schistosoma* spp.	X	X	X	X		X	X	X		X		X	X					X	X	
Nematodes	Anisakiasis	*Anisakis* spp.					X								X					X		
	Ancylostomiasis, cutaneous larval migrans	*Ancylostoma* spp., *Bunostomum* spp., *Uncinaria* spp.		X	X	X															X	
	Ascariasis, visceral larval migrans	*Toxocara* spp., *Ascaris* spp., *Toxascaris* spp., *Amplicaecum* spp., *Hexametra* spp., *Lagochilascaris* spp., *Neoascaris* spp., *Parascaris* spp.	X	X	X	X	X		X	X		X		X	X					X		
	Capillariasis	*Capillaria* spp.		X		X	X			X		X			X					X		
	Dioctophyma infection	*Dioctophyma renale*		X	X	X	X		X					X	X					X		
	Dracunculiasis	*Dracunculus* spp.				X									X					X		

Nematodes (continued)	Filariasis	*Brugia* spp., *Dipetalonema* spp., *Dirofilaria* spp., *Loa loa*, *Onchocerca* spp., *Wucheraria* spp.		X		X				X				X	X			
	Gnathostomiasis	*Gnathostoma* spp.		X	X	X							X	X			X	X
	Parasitic meningo-encephalitis	*Angiostrongylus* spp.					X				X						X	
	Strongyloidiasis	*Strongyloides* spp.		X	X	X			X	X	X	X	X	X			X	X
	Thelaziasis[d]	*Thelazia* spp.		X		X						X			X			
	Trichinosis	*Trichinella spiralis*		X		X					X		X	X			X	
	Trichostrongyliasis	*Trichostrongylus* spp.			X			X	X			X					X	
Cestodes	Dog tapeworm	*Dipylidium caninum*		X		X											X	
	Dwarf tapeworm	*Hymenolepis nana*				X				X	X						X	
	Fish tapeworm	*Diphyllobothrium latum*		X		X	X						X	X			X	
	Hydatidosis	*Echinococcus* spp.		X	X	X		X			X	X	X	X			X	
	Mouse/rat tapeworm	*Hymenolepis diminuta*			X				X		X						X	
	Raillietiniasis[d]	*Raillietina* spp.	X							X	X						?	
	Sparagnosis	*Spirometra* spp., *Diphyllobothrium* spp.	X	X						X	X		X	X		X	X	X
	Tapeworm, coenuriasis	*Taenia* spp.			X	X					X	X	X	X			X	

Continued

Microorganism Group	Disease	Agent	Animal Source[a] A	F	B	D	P	C	E	N	L	M	O	S	W	Route of Infection[b] Ae	Ab	Bs	Co	In	Sp	Wi
Protozoa	Amoebiasis	*Entamoeba histolytica*				X				X										X		
	Babesiosis, piroplasmosis	*Babesia* spp.			X			X				X	X		X		X					
	Balantidiasis	*Balantidium coli*								X				X						X		
	Coccidiosis	*Isospora* spp.		X		X														X		
	Giardiasis	*Giardia lamblii*				X				X		X	X	X	X					X		
	Leishmaniasis	*Leishmania* spp.		X		X			X			X	X		X		X					
	Malaria	*Plasmodium* spp.								X							X					
	Naegleriosis	*Naegleria fowlerii*	X		X							X				X				X		
	Pneumocystis infection	*Pneumocystis carinii*		X	X	X					X	X				?			?			
	Sarcosporidiosis	*Sarcocystis* spp.	X		X			X	X				X	X						X		
	Toxoplasmosis[d]	*Toxoplasma gondii*	X	X	X	X		X				X	X	X						X		
	Trypanosomiasis	*Trypanosoma* spp.		X	X	X		X	X	X	X	X	X	X	X		X	X			X	
RNA viruses (Flavivirus)	Banzi[d]	BAN virus										X					X					
	Dengue	DEN virus								X							X					
	Ilheus[d]	ILH virus	X														X					
	Japanese encephalitis	JBE virus	X		X				X					X			X					
	Kyasanur forest disease	KFD virus	X							X		X			X		X					
	Louping ill	LI virus	X					X				X	X				X					

RNA viruses (Flavivirus, continued)	Murray Valley encephalitis	MVE virus	X										X		
	Powassan encephalitis	POW virus							X				X		
	Rio Bravo infection, Bat salivary gland virus fever	RB virus									X	?		X	
	Russian spring-summer encephalitis, Central European encephalitis, Tick-borne encephalitis	Hypr virus, Kumlinge virus, Omsk virus, RSSE virus	X		X	X			X	X	X		X		
	St. Louis encephalitis	SLE virus	X								X		X		
	Wesselsbron	WSL virus			X	X				X			X		
	West Nile	WN virus	X										X		
	Yellow fever	YF virus						X					X		X
	Zika	Zika virus						X					X		
RNA viruses (Togavirus)	Chikungunya	CHIK virus						X					X		
	Eastern encephalitis	EEE virus	X				X					X	X		
	Mayaro	MAY virus	X										X		
	O'nyong-nyong	ONN virus						X					X		
	Semliki Forest encephalitis	SF virus							X			X			
	Sindbis	SIN virus		X									X		

Continued

Microorganism Group	Disease	Agent	Animal Source[a]													Route of Infection[b]						
			A	F	B	D	P	C	E	N	L	M	O	S	W	Ae	Ab	Bs	Co	In	Sp	Wi
RNA viruses (Togavirus, continued)	Venezuelan encephalitis	VEE virus	X						X			X				X	X					
	Western encephalitis	WEE virus	X						X							X	X					
RNA viruses (Bunyavirus)	Apeu	Apeu virus								X		X					X					
	California encephalitis	CE virus									X	X			X		X					
	Caraparu	CAR virus								X		X					X					
	Catu	Catu virus								?					X		X					
	Congo-Crimean hemorrhagic fever	CCHF virus	X		X			X			X	X	X				X					
	Germiston	GER virus										X					X					
	Guama[d]	GMA virus								?					X		X					
	Hemorrhagic fever	Hantaan virus										X				X			X	X		
	Itaqui[d]	ITQ virus								X		X					X					
	LaCrosse encephalitis	LAC virus									X	X			X		X					
	Madrid[d]	MAD virus										X					X					
	Marituba	MTB virus								X		X					X					
	Murutucu	MUR virus								X		X					X					
	Nairobi sheep disease (Dugbe fever, Ganjam fever)	NSD virus (DUG virus, GAN virus)			X			X					X				X					
	Oriboca	ORI virus								X		X					X					
	Oropouche	ORO virus								?					X		X					

RNA viruses (Bunyavirus, continued)	Rift Valley fever	RVF virus or Zinga virus			X			X					X		X		X	X		X			
	Tahyna	TAH virus									X	X						X					
RNA viruses (Orbivirus)	Colorado tick fever	CTF virus										X			X			X					
	Kemerovo fever[d]	KEM virus			X													X					
RNA viruses (Rhabdovirus)	Rabies	Rabies virus		X	X	X		X	X				X	X	X		X		X				X
	Vesicular stomatitis	VS virus			X				X			X		X				X		X			
RNA viruses (Reovirus)	Reovirus infection[d]	Reovirus		?	?	?				?		?	?	?	?						?		
	Rotavirus infection[d]	Rotavirus								?											?		
RNA viruses (Picornavirus)	Coxsackie disease[d]	Coxsackie virus					X					X		X			X			X			
	Foot and mouth disease	FMD virus			X			X					X	X	X		X			X			
	Hepatitis	Hepatitis A virus								X							X			X	X		
RNA viruses (Arenavirus)	Lymphocytic choriomeningitis	LCM virus				X				X		X		X						X	X		
	Rodent-borne hemorrhagic fevers	Junin virus, Lassa fever virus, Machupo virus, Tacaribe virus										X			X		X			X	X		
RNA viruses (Others)	Influenza	Influenza virus	X			X			X			X		X			X			X			
	Green monkey disease, African hemorrhagic fever	Ebola virus, Marburg virus								X										X			X

Continued

Microorganism Group	Disease	Agent	Animal Source[a]													Route of Infection[b]						
			A	F	B	D	P	C	E	N	L	M	O	S	W	Ae	Ab	Bs	Co	In	Sp	Wi
RNA viruses (Others, continued)	Newcastle disease	ND virus	X													X			X			
	Simian hepatitis	Simian hepatitis virus								X									X	X		
DNA viruses	Contagious ecthyma, Bovine papular stomatitis, Pseudocowpox	Orf virus			X			X					X						X			
	Pox	Cowpox virus, monkeypox virus, Yaba virus, (Catpox virus, Tanapox virus)		X	X					X		X			X			X	X			X
	Pseudorabies	Pseudorabies virus												X		?			X			
	Simian herpes	Simian herpesvirus B								X						?		X	X	?		X
Unclassified	Cat scratch disease	Unknown		X		X						X						X				X

[a]A = birds, fowl, poultry (avian); F = cats (felines); B = cattle (bovines); D = dogs (canines); P = fish, seafood (piscines); C = goats (caprines); E = horses (equines); N = nonhuman primates; L = rabbits (lagomorphs); M = rodents (murines); O = sheep (ovines); S = swine (porcines); W = wildlife including bats, marsupials, and other ruminants.

[b]Ae = aerosol; Ab = arthropod bite; Bs = bite/scratch; Co = contact; In = ingestion; Sp = skin penetration; Wi = wound infection.

[c]Fetal infection, *in utero*.

[d]Epidemiologic relationship of animals to human disease has not been proven.

SOURCE: The information for this table was derived from references 6, 17, 21, 26, 53, and 69.

Appendix E

Regulations Governing the Packaging, Labeling, and Transport of Infectious Agents

CONTENTS

The importation and interstate transport of infectious agents is subject to several federal regulations. In this appendix, most of the relevant regulations and forms are reproduced for the information of those who will use this report. Federal regulations and forms are subject to revision, and the reader is urged to consult the most up-to-date version of the *Code of Federal Regulations*, as well as relevant offices in the U.S. Department of Agriculture and the Public Health Service, to remain abreast of changes that may have occurred.

EXCERPT FROM *CODE OF FEDERAL REGULATIONS*, TITLE 9, PART 122—ORGANISMS AND VECTORS*

PART 122—ORGANISMS AND VECTORS

AUTHORITY: Sec. 2, 32 Stat. 792, 37 Stat. 832-833; 21 U.S.C. 111, 151-158.

EDITORIAL NOTE: For nomenclature changes, see 36 FR 24928, Dec. 24, 1971.

§ 122.1 Definitions.

The following words, when used in the regulations in this Part 122, shall be construed, respectively, to mean:

(a) *Department.* The U.S. Department of Agriculture.

(b) *Secretary.* "Secretary" means the Secretary of Agriculture of the United States, or any officer or employee of the Department to whom authority has heretofore been delegated, or to whom authority may hereafter be delegated, to act in his stead.

(c) *Veterinary Services.* The Veterinary Services unit of the Department.

(d) *Deputy Administrator.* The Deputy Administrator, Veterinary Services or any officer or employee of the Veterinary Services to whom authority has heretofore lawfully been delegated, or may hereafter lawfully be delegated, to act in his stead.

(e) *Organisms.* All cultures or collections of organisms or their derivatives, which may introduce or disseminate any contagious or infectious disease of animals (including poultry).

(f) *Vectors.* All animals (including poultry) such as mice, pigeons, guinea pigs, rats, ferrets, rabbits, chickens, dogs, and the like, which have been treated or inoculated with organisms, or which are diseased or infected with any contagious, infectious, or communicable disease of animals or poultry or which have been exposed to any such disease.

(g) *Permittee.* A person who resides in the United States or operates a business establishment within the United States, to whom a permit to import or transport organisms or vectors has been issued under the regulations.

(h) *Person.* Any individual, firm, partnership, corporation, company, society, association, or other organized group of any of the foregoing, or any agent, officer, or employee of any thereof.

[31 FR 81, Jan. 5, 1966]

§ 122.2 Permits required.

No organisms or vectors shall be imported into the United States or transported from one State or Territory or the District of Columbia to another State or Territory or the District of Columbia without a permit issued by the Secretary and in compliance with the terms thereof: *Provided,* That no permit shall be required under this section for importation of organisms for which an import permit has been issued pursuant to Part 102 of this subchapter or for transportation of organisms produced at establishments licensed under Part 102 of this subchapter. As a condition of issuance of permits under this section, the permittee shall agree in writing to observe the safeguards prescribed by the Deputy Administrator for public protection with respect to the particular importation or transportation.

(Approved by the Office of Management and Budget under control number 0579-0013)

[28 FR 7896, Aug. 2, 1963. Redesignated at 31 FR 81, Jan. 5, 1966 and amended at 48 FR 57473, Dec. 30, 1983]

§ 122.3 Application for permits.

The Secretary may issue, at his discretion, a permit as specified in § 122.2 when proper safeguards are set up as provided in § 122.2 to protect the public. Application for such a permit shall be made in advance of shipment, and each permit shall specify the name and address of the consignee, the true name and character of each of the organisms or vectors involved, and the use to which each will be put.

(Approved by the Office of Management and Budget under control number 0579-0015)

[23 FR 10065, Dec. 23, 1958. Redesignated at 31 FR 81, Jan. 5, 1966 and amended at 48 FR 57473, Dec. 30, 1983]

§ 122.4 Suspension or revocation of permits.

(a) Any permit for the importation or transportation of organisms or vectors issued under this part may be formally suspended or revoked after opportunity for hearing has been accorded the permittee, as provided in Part 123 of this subchapter, if the Secretary finds that the permittee has failed to observe the safeguards and instructions prescribed by the Deputy Administrator with respect to the particular importation or transportation or that such importation or transportation for any other reason may result in the introduction or dissemination from a foreign country into the United States, or from one State, Territory or the District of Columbia to another, of the contagion of any contagious, infectious or communicable disease of animals (including poultry).

(b) In cases of wilfulness or where the public health, interest or safety so requires, however, the Secretary may without hearing informally suspend such a permit upon the grounds set forth in paragraph (a) of this section, pending determination of formal proceedings under Part 123 of this subchapter for suspension or revocation of the permit.

[23 FR 10065, Dec. 23, 1958. Redesignated at 31 FR 81, Jan. 5, 1966]

**Code of Federal Regulations*. 1988. Part 122, Organisms and Vectors, in Chapter 1, Animal and Plant Health Inspection Service.

STATEMENT FROM USDA'S ANIMAL AND PLANT HEALTH AND INSPECTION SERVICE ON REQUIREMENTS FOR IMPORTING CELL CULTURES**

IMPORTING CELL CULTURES: REQUIREMENTS TO PROTECT U.S. AGRICULTURE

Some researchers don't know it—and others have found out the hard way—but they need a permit from the U.S. Department of Agriculture to bring cell cultures into the United States. USDA's Animal and Plant Health Inspection Service requires permits for importing cells or their culture medium because of the possibility they might carry foreign animal diseases that could devastate a highly susceptible U.S. livestock population.

Cell cultures and similar materials arriving in this country without permits are confiscated and destroyed by agricultural inspectors at U.S. ports of entry. Inspectors have scientific backgrounds and don't like having to set back research. But at that point there is no alternative. Protection against introduction of organisms from abroad must be built into the import procedure ahead of time.

A major concern is foot-and-mouth disease (FMD), which is found throughout most of the world. Two of the six FMD outbreaks that occurred in the United States early in this century were traced to contaminated cowpox vaccine. FMD has not occurred in the United States since 1929, when stringent import laws were enacted.

Although scientists have exchanged cell lines for many years, there is a large increase in this activity. One reason is the great popularity of monoclonal antibodies as research tools because of their high specificity and reproducibility. Monoclonal antibodies are produced by hybridomas, which are lines of hybrid cells formed by fusing mammalian cells. Fetal calf serum, a possible source of FMD virus, is used to produce virtually all such cultures.

Before most cell lines can be imported, safety tests must be conducted at the APHIS Foreign Animal Disease Diagnostic Laboratory at Plum Island, N.Y., a high-security facility located off the northeastern tip of Long Island. Generally, the tests employ recently developed *in vitro* methods, which are considerably faster and less expensive than previous safety tests conducted in living animals. However, tests for FMD still require the use of live animals.

To get an import permit, complete the questionnaire "Importation Information" and fill out VS Form 16-3. Send both documents to:

> Organisms and Vectors Section
> Import-Export and Emergency Planning Staff
> VS-APHIS-USDA
> 6505 Belcrest Road
> Hyattsville, MD 20782
> (Phone: 301-436-5453)

Applicants will be notified if a safety test is required. If so, they will be told the estimated cost, and they must then deposit funds to cover this cost. Importers bringing in cell cultures or hybridoma cells on a regular basis can establish an escrow account, which avoids the need to forward funds for each import. A *minimum* of four vials, each containing at least 1 million cells from a uniform lot, is required for a safety test.

The normal working time for issuing a permit, completing safety tests, and transferring the imported material is 60 to 90 days. To expedite the procedure, APHIS may issue a permit for the material to be shipped to the Foreign Animal Disease Diagnostic Laboratory at Plum Island, N.Y., pending receipt of the funds. However, actual testing will not begin until the funds are in hand. Cost of testing varies. In early 1986, an *in vivo* safety test using susceptible

**Reprinted from *APHIS Facts*, June 1986. Washington, DC: U.S. Department of Agriculture, Animal and Plant Health Inspection Service.

host animals cost from $2,000 to $3,000. However, it sometimes was possible to cut costs by pooling samples for host animal tests. Cost of *in vitro* safety tests was about $500 per test, depending on the animal diseases present in the country of origin and the intended use of the material being imported.

Safety testing may not be required for some cell cultures imported for human diagnostic or research purposes. Some examples are cultured human bone marrow cells, amniocentesis samples, or cells to be karyotyped.

Official information on Federal requirements for importing cell cultures (including hybridomas) is published in Veterinary Services Memorandum 593.1, March 11, 1986, which is available from the Organisms and Vectors Section. This group can also provide additional details and answer questions from prospective importers.

Researchers should plan their imports as far in advance as possible. That way, the actual import will proceed smoothly and expeditiously when the time comes.

VETERINARY SERVICES MARCH 1986 MEMORANDUM ON IMPORTATION OF CELL CULTURES INCLUDING HYBRIDOMAS

Veterinary Services Memorandum 593.1

Subject: Importation of Cell Cultures Including Hybridomas

To: Area Veterinarians in Charge, VS
Directors, VS Regions
Veterinary Medical Officers, VS
Veterinary Medical Officers, PPQ
Director, National Veterinary Services Laboratories
Director, National Program Planning Staffs
Chief, Foreign Animal Disease Diagnostic Laboratory

I PURPOSE

The purpose of this memorandum is to provide updated information on importing cell cultures, including hybridomas, that may require safety testing.

II CANCELLATIONS

This memorandum replaces VS Notice dated June 25, 1985.

III POLICY

It is Veterinary Services policy that no animal-origin biological materials such as cell cultures, monoclonal antibodies, or related material may be imported into the United States without a Veterinary Services (VS) permit (VS Form 16-3A).

IV GENERAL

A. To obtain an import permit, an application (VS Form 16-3) should be submitted to:

Import-Export Staff, Organisms and Vectors
VS, APHIS, USDA
6505 Belcrest Road
Hyattsville, MD 20782

A sample copy of the application (Form 16-3) is enclosed (Attachment 1). Applicants must also complete the questionnaire entitled "Importation Information" (Attachment 2) and submit it with their application.

B. The information requested in these forms is necessary for proper evaluation of the request. Incomplete information will result in denials or delays in processing the application. Based upon the information submitted by the applicant, a determination will be made if the material to be imported requires safety testing to ensure it is free from livestock pathogens. Safety testing can be conducted at the Foreign Animal Disease Diagnostic Laboratory (FADDL), Plum Island, New York.

C. Applicants will be advised if a safety test is required and will be given an estimate of the cost for conducting the test. Applicants desiring to have material safety tested must enter into a Cooperative Trust Fund Agreement with APHIS, VS, and deposit sufficient funds to cover the estimated cost for safety testing in advance. The Import-Export Animals and Products Staff will initiate the Cooperative Trust Fund Agreement. In order to expedite the procedure, VS may issue a permit for the material to be shipped to FADDL pending receipt of the funds and Cooperative Trust Fund Agreement. However, the signed Cooperative Agreement, plus the necessary funds, must have been received by VS before testing can be scheduled at FADDL.

D. The normal working time for issuing a permit for importing material to Plum Island, New York, completing safety tests, and transferring the imported material to the applicant is 60 to 90 days. A *minimum* of four vials, each containing at least one million cells from a uniform lot, is required for a safety test.

E. Once the safety test is completed and a determination made that the imported material is free from livestock pathogens, the remainder of the imported material will be released directly to the importer under conditions specified in the permit (Attachment 3).

F. If an importer wishes to import cell cultures and/or hybridoma cells on a regular basis, the applicant may enter into a continuous Cooperative Trust Fund Agreement with VS and establish an escrow account to ensure no unnecessary delay will occur because of lack of funds (Attachment 4).

G. Presently, *in vivo* safety tests utilizing susceptible host animals usually cost approximately $2,000 to $3,000 per test. Sometimes it is possible to reduce the cost by pooling samples in one host animal test. Scientists at FADDL have developed *in vitro* safety tests to detect certain livestock pathogens resulting in substantial cost savings for importers. *In vitro* safety tests for additional diseases are being developed. Current cost for *in vitro* tests is approximately $500 per test depending upon animal disease present in the country of origin and the intended use of material being imported.

H. Safety testing may not be required for some cell cultures imported for human diagnostic purposes and research. Examples of material which would qualify without safety testing include cultured human bone marrow cells, amniocentesis samples, or cells to be karyotyped.

Applications for such cell cultures will be individually evaluated.

I. The following classification of cell cultures is based on intended use and generally indicates the level of safety testing required.

Class I — Cell cultures to be used for the production of products such as vaccines, hormones, or other biologicals to be used in livestock, poultry, or for commercial distribution.

Requirement: These cell cultures must be safety tested at FDDL using susceptible host animals, approved in vitro test, and/or laboratory animals.

Class II — Cell cultures to be used only for in vitro studies and not to be used in animals other than primates.

Requirements: These cultures may not require safety testing. The material may be sent directly to the importer when no safety testing is required. The permit (VS form 16-3A) will specify restrictions such as "FOR *IN VITRO* LABORATORY TESTS: DO NOT INOCULATE INTO LIVESTOCK, BIRDS, OR LABORATORY ANIMALS."

J. Cell cultures imported under permit which do not require a safety test may not be distributed to other laboratories without prior approval from USDA, APHIS, VS. Applications for the distribution of imported material should be submitted to the USDA, APHIS, VS, Import-Export Staff, Organisms and Vectors.

J.K. Atwell
Deputy Administrator
Veterinary Services

4 Enclosures

Attachment 1—VS Form 16-3
Application for Permit to Import Controlled Material or to Import or Transport Organisms or Vectors

No controlled material, organisms or vectors may be imported or moved interstate unless the data requested on this form is furnished and certified (9 CFR 94, 95, and 122).

U S DEPARTMENT OF AGRICULTURE
ANIMAL AND PLANT HEALTH INSPECTION SERVICE
VETERINARY SERVICES
FEDERAL BUILDING, HYATTSVILLE, MARYLAND 20782

FORM APPROVED - OMB NO 0579-0015

APPLICATION FOR PERMIT TO:

☐ **IMPORT CONTROLLED MATERIAL**

☐ **IMPORT OR TRANSPORT ORGANISMS OR VECTORS**

1 MODE OF TRANSPORTATION

2 U S PORT(S) OF ENTRY

INSTRUCTIONS: Submit 2 copies to address above. Attach additional sheets, if necessary.

3. TO: *(Name, address, and phone no. of applicant - include Zip Code)*

4. FROM: *(Name and address of shipper - include Zip Code)*

5 DESCRIPTION OF MATERIAL *(Name of material, country of origin, animal source, etc.)*

6 QUANTITY OF MATERIAL TO BE IMPORTED AND FREQUENCY OF IMPORTATIONS

7 PROPOSED USE OF MATERIAL, EXPECTED COMPLETION DATE, AND FINAL DISPOSITION TO BE MADE

8 DESCRIPTION OF APPLICANT'S FACILITIES AND EQUIPMENT FOR HANDLING MATERIAL

9 QUALIFICATIONS OF TECHNICAL PERSONNEL WHO WILL BE WORKING WITH THIS MATERIAL

10 METHOD OF TREATMENT OF MATERIAL *(Disease Safeguard)*

11 WORK OBJECTIVES, PROPOSED PLAN OR WORK, AND ADDITIONAL PERTINENT INFORMATION

12 PERTINENT PUBLISHED PAPER OR ABSTRACT *(Please attach copy, if available)*

CHECK IF COPY IS ATTACHED ☐

I CERTIFY THIS MATERIAL WILL BE USED IN ACCORDANCE WITH ALL RESTRICITONS AND PRECAUTIONS AS MAY BE SPECIFIED IN THE PERMIT.

13 SIGNATURE OF APPLICANT	14 TYPED NAME OF OFFICIAL SIGNING
15 DATE SIGNED	16 TYPED TITLE OF OFFICIAL SIGNING

VS FORM 16-3 (JAN 88) (Previous editions are obsolete.)

Attachment 2—
Importation Information Questionnaire for
Cell Lines

Cell line designation, or reference number: ______________________

Country of origin of cell line: ______________________

Cell line passage history: ______________________

Country of origin of culture media: ______________________

Type of culture media: ______________________

Source of culture media (   ): ______________________

Country of origin <u>and</u> source of any nutritive factors of animal origin in the culture media (e.g. serum or supplements): ______________________

____

If serum is used, indicate percentage: ______________________

If serum is used, and it is of USA origin, is a USDA Export Certificate available: Yes _____ No _____. If answer is yes, give company's name and serum lot number ______________________

Country of origin <u>and</u> source of any animal enzymes (e.g. trypsin) which have been used to cultivate the cells? ______________________

____

Country of origin <u>and</u> source of any animal viruses utilized in the laboratory where the cell line originates? ______________________

____

If cell line is a hybridoma, specify fusion partners: ______________________

____

If cell line is <u>not</u> a hybridoma, specify its origin or derivation (e.g. EBV transformation of human B lymphocytes): ______________________

____

Potential use of imported material: ______________________

Name and address of the institution where the material originated: __________

____

____

______________________________________ ______________________
Signature of Authorized Company Representative Date

Attachment 3—VS Form 16-3A
Veterinary Permit, International and Domestic Control Organisms and Vectors

☆ U.S. GOVERNMENT PRINTING OFFICE: 1987—190-784

UNITED STATES DEPARTMENT OF AGRICULTURE
ANIMAL AND PLANT HEALTH INSPECTION SERVICE
VETERINARY SERVICES

VETERINARY PERMIT
INTERNATIONAL AND DOMESTIC CONTROL
ORGANISMS AND VECTORS

PERMIT NO.

DATE ISSUED

DATE EXPIRES *(For purposes of movement from shipper to permittee)*

NAME AND ADDRESS OF SHIPPER

TO *(Name and Address of Permittee)*

MODE OF TRANSPORTATION

U. S. PORT OF ARRIVAL

AS REQUESTED IN YOUR APPLICATION YOU ARE AUTHORIZED TO IMPORT OR TRANSPORT THE FOLLOWING MATERIALS

RESTRICTIONS AND PRECAUTIONS FOR TRANSPORTING AND HANDLING MATERIALS AND ALL THEIR DERIVATIVES
(Item 1 is always applicable and Items 2 through 11 are applicable only when "X".)

This permit is issued under authority contained in Parts 94 and 122, Chapter 1, Title 9, CFR.
The authorized materials or their derivatives shall be used only in accordance with the restrictions and precautions specified below.
(ALTERATIONS OF RESTRICTIONS CAN BE MADE ONLY WHEN AUTHORIZED BY VETERINARY SERVICES.)

1. Adequate safety precautions shall be maintained during shipment and handling to prevent dissemination of disease.
2. ☐ Work shall be limited to **IN VITRO** laboratory studies only.
3. ☐ **This permit does not authorize direct or indirect exposure of domestic animals, including poultry, cattle, sheep, swine, horses, etc.**
4. ☐ All animals shall be exposed and held only in isolated insect and rodent-proof facilities.
5. ☐ All equipment, animals, pens, cages, bedding, waste, etc. in direct or indirect contact with these materials shall be sterilized by autoclaving or incineration.
6. ☐ Packaging materials, containers, and all unused portions of the imported materials shall be sterilized by autoclaving or incineration.
7. ☐ Materials shall be shipped by Registered Mail, Railway or Air Express.
8. ☐ Acknowledge receipt of materials by completing and mailing attached VS Form 16-19, which requires no stamp.
9. ☐ This permit is valid only for work conducted or directed by you in your present facilities. *(MATERIALS SHALL NOT BE REMOVED TO ANOTHER LOCATION, NOR DISTRIBUTED TO OTHERS, WITHOUT USDA AUTHORIZATION,)*
10. ☐ On completion of your work, all permitted materials and all derivatives therefrom shall be destroyed and the USDA must be promptly notified by completing and mailing the attached VS Form 16-20, which requires no stamp.
11. ☐

To expedite clearances at the Port of Entry the shipper should attach one of the enclosed labels to each package of authorized material. *(Additional labels may be obtained from this office.)*

NO. LABELS ENCLOSED

SIGNATURE

TITLE

VS FORM 16-3A *Previous editions obsolete.*
(MAR 73)

Attachment 4—
Procedure to Complete Trust Fund Agreement for Safety Testing of Cell Cultures, Including Hybridomas

When the applicant has been advised by the Import-Export Animals and Products Staff that a safety test is required (see VS Memorandum No. 593.1, dated March 11, 1986), the following steps will be taken to obtain the Cooperative Trust Fund Agreement:

1. A copy of the application with other pertinent information will be submitted to FADDL, Plum Island, New York, to obtain cost estimates for the required safety tests.

2. The total cost estimate will be sent to the applicant.

3. The Import-Export Staff will then draft the Cooperative Trust Fund Agreement.

4. The VS Executive Office will review the draft document for policy compliance, proper format, and funding requirements.

5. The Import-Export Staff will finalize and send five copies of the Cooperative Trust Fund Agreement to the Cooperator for signature.

6. If the agreement is acceptable, the Cooperator will sign all copies and return the agreement to the Import-Export Staff with a check for the total cost of the tests made payable to the United States Department of Agriculture, Animal and Plant Health Inspection Service, Veterinary Services.

7. The Import-Export staff will deliver the check and Cooperative Trust Fund Agreement to the VS Executive Office for a document number, final signature and processing.

8. Two copies of the executed Cooperative Trust Fund Agreement will be sent to the Cooperator for their records.

9. If an importer wishes to import cell cultures and/or hybridoma cells on a regular basis, the importer should inform the Import-Export Staff during initial discussions. The importer may enter into a continuous Cooperative Trust Fund Agreement with VS and establish a revolving trust fund account.

VS Form 16-18
Authorized Entry Label

PERISHABLE ANIMAL QUARANTINE MATERIAL

AUTHORIZED ENTRY BY

U.S. DEPARTMENT OF AGRICULTURE
ANIMAL AND PLANT HEALTH INSPECTION SERVICE
VETERINARY SERVICES
FEDERAL BUILDING
HYATTSVILLE, MARYLAND 20782

PLEASE EXPEDITE

VETERINARY PERMIT NO.	EXPIRES

THIS PACKAGE CONTAINS

VS FORM 16-18
(MAR 74)

EXCERPT FROM, AND INSTRUCTIONS BASED ON, *CODE OF FEDERAL REGULATIONS*, TITLE 42, PART 71— FOREIGN QUARANTINE: IMPORTATION OF ETIOLOGICAL AGENTS, HOSTS, AND VECTORS

UNITED STATES DEPARTMENT OF HEALTH AND HUMAN SERVICES
Public Health Service

42 CFR - Part 71
Foreign Quarantine

Importation of Etiological Agents, Hosts, and Vectors

§71.54. Etiological Agents, Hosts, and Vectors

(a) A person may not import into the United States, nor distribute after importation, any etiological agent or any arthropod or other animal host or vector of human disease, or any exotic living arthropod or other animal capable of being a host or vector of human disease unless accompanied by a permit issued by the Director.

(b) Any import coming within the provisions of this section will not be released from custody prior to receipt by the District Director of the U.S. Customs Service of a permit issued by the Director.

INSTRUCTIONS

1. *Classes of Imports Requiring Permits.* It is impracticable to list all of the several hundred species of etiological agents and vectors that may be covered by §71.54. Certain classes of imports over which the maintenance of surveillance is important and for which permits must be obtained from the Director, Centers for Disease Control, Public Health Service, Department of Health and Human Services, or his/her authorized representative, follow:

a. Any living insect, or other living arthropod, known to be or suspected of being infected with any disease transmissible to man; also, if alive, any bedbugs, fleas, flies, lice, mites, mosquitoes, or ticks, even if uninfected. This includes eggs, larvae, pupae, and nymphs as well as adult forms.

b. Any animal known to be or suspected of being infected with any disease transmissible to man.

c. *All* live bats.

d. Unsterilized specimens of human and animal tissue (including blood), body discharges or excretions, or similar material, when known to be or suspected of being infected with disease transmissible to man.

e. Any culture of living bacteria, virus, or similar organism known to cause or suspected of causing human diseases.

f. Any snails capable of transmitting schistosomiasis. *No mollusks* are to be admitted without a permit from either the Public Health Service or the Department of Agriculture. Any shipment of mollusks with a permit from either agency should be cleared immediately.

2. *Advice to Customs.* In applying this section, Customs officers may request advice of the nearest quarantine office or Program headquarters if a question should arise as to the necessity for a permit in any individual instance. In giving advice, quarantine officers should be guided by the principle that the intent of this section is to keep out of the United States communicable diseases and also vectors or hosts not commonly found in this country. When an importation does not seem likely to bring in disease or a vector or host, advice should be given to the effect that a permit is not required. In case of doubt concerning admissibility, quarantine officers should promptly make inquiry of Program headquarters.

3. *Blanket Permits.* Blanket permits, limited as to time and material, are occasionally issued. The original blanket permit is retained in the office to which it is issued. When it is used as authority to import quarantinable material, a certified or photostatic copy of it should accompany the shipment. The copy is cancelled and collected as in the case of a single entry permit.

4. To obtain an importation permit, one should complete the attached permit application form and send to: Office of Biosafety, Centers for Disease Control, Atlanta, Georgia 30333. A permit is issued to the recipient, and may be conditional upon one or more of the items indicated on the attached sample permit.

Importation or Transfer Authorization
Label (CDC 0.1007)

IMPORTATION OR TRANSFER AUTHORIZED **BY**

PHS Permit No. ______________________

Expiration Date ______________________

TO:

DO NOT OPEN IN TRANSIT

BIOMEDICAL MATERIALS
ETIOLOGICAL AGENTS OR VECTORS

NOTICE TO CARRIER: If inspection on arrival in U.S. reveals evidence of damage or leakage, immediately notify: Director, Centers for Disease Control, Atlanta, Georgia 30333 – Telephone 404–633–5313. CDC 0.1007 6/85

Shipper's Declaration for Dangerous Goods (Sample Form)

SHIPPER'S DECLARATION FOR DANGEROUS GOODS

Shipper

Air Waybill No.

Page of Pages

Shipper's Reference Number
(optional)

Consignee

Two completed and signed copies of this Declaration must be handed to the operator

WARNING

Failure to comply in all respects with the applicable Dangerous Goods Regulations may be in breach of the applicable law, subject to legal penalties. This Declaration must not, in any circumstances, be completed and/or signed by a consolidator, a forwarder or an IATA cargo agent.

TRANSPORT DETAILS

This shipment is within the limitations prescribed for:
(delete non-applicable)

PASSENGER AND CARGO AIRCRAFT	CARGO AIRCRAFT ONLY

Airport of Departure

Airport of Destination:

Shipment type: *(delete non-applicable)*

NON-RADIOACTIVE	RADIOACTIVE

NATURE AND QUANTITY OF DANGEROUS GOODS

Dangerous Goods Identification				Quantity and type of packing	Packing Inst.	Authorization
Proper Shipping Name	Class or Division	UN or ID No.	Subsidiary Risk			

Additional Handling Information

I hereby declare that the contents of this consignment are fully and accurately described above by proper shipping name and are classified, packed, marked and labelled, and are in all respects in the proper condition for transport by air according to the applicable International and National Government Regulations.

Name/Title of Signatory

Place and Date

Signature
(see warning above)

STYLE F83 LABELMASTER, CHICAGO, IL 60646

EXCERPT FROM *CODE OF FEDERAL REGULATIONS*, TITLE 42, PART 72— INTERSTATE SHIPMENT OF ETIOLOGIC AGENTS

PART 72—INTERSTATE SHIPMENT OF ETIOLOGIC AGENTS [1]

Sec.
72.1 Definitions.
72.2 Transportation of diagnostic specimens, biological products, and other materials; minimum packaging requirements.
72.3 Transportation of materials containing certain etiologic agents; minimum packaging requirements.
72.4 Notice of delivery; failure to receive.
72.5 Requirements; variations.

Authority: Sec. 215, 58 Stat. 690, as amended, 42 U.S.C. 216; sec. 361, 58 Stat. 703, (42 U.S.C. 264)

§ 72.1 Definitions.

As used in this part:

"Biological product" means a biological product prepared and manufactured in accordance with the provisions of 9 CFR Parts 102–104 and 21 CFR Parts 312 and 600–680 and which, in accordance with such provisions, may be shipped in interstate traffic.

"Diagnostic specimen" means any human or animal material including, but not limited to, excreta, secreta, blood and its components, tissue, and tissue fluids being shipped for purposes of diagnosis.

"Etiologic agent" means a viable microorganism or its toxin which causes, or may cause, human disease.

"Interstate traffic" means the movement of any conveyance or the transportation of persons or property, including any portion of such movement or transportation which is entirely within a State or possession, (a) from a point of origin in any State or possession to a point of destination in any other State or possession, or (b) between a point of origin and a point of destination in the same State or possession but through any other State, possession, or contiguous foreign country.

[1] The requirements of this part are in addition to and not in lieu of any other packaging or other requirements for the transportation of etiologic agents in interstate traffic prescribed by the Department of Transportation and other agencies of the Federal Government.

§ 72.2 Transportation of diagnostic specimens, biological products, and other materials; minimum packaging requirements.

No person may knowingly transport or cause to be transported in interstate traffic, directly or indirectly, any material including, but not limited to, diagnostic specimens and biological products which such person reasonably believes may contain an etiologic agent unless such material is packaged to withstand leakage of contents, shocks, pressure changes, and other conditions incident to ordinary handling in transportation.

§ 72.3 Transportation of materials containing certain etiologic agents; minimum packaging requirements.

Notwithstanding the provisions of § 72.2, no person may knowingly transport or cause to be transported in interstate traffic, directly or indirectly, any material (other than biological products) known to contain, or reasonably believed by such person to contain, one or more of the following etiologic agents unless such material is packaged, labeled, and shipped in accordance with the requirements specified in paragraphs (a)–(f) of this section:

Bacterial Agents

Acinetobacter calcoaceticus.
Actinobacillus—all species.
Actinomycetaceae—all members.
Aeromonas hydrophila.
Arachnia propionica.
Arizona hinshawii—all serotypes.
Bacillus anthracis.
Bacteroides spp.
Bartonella—all species.
Bordetella—all species.
Borrelia recurrentis, B. vincenti.
Brucella—all species.
Campylobacter (Vibrio) foetus, C. (Vibrio) jejuni.
Chlamydia psittaci, C. trachomatis.
Clostridium botulinum, Cl. chauvoei, Cl. haemolyticum, Cl. histolyticum, Cl. novyi, Cl. septicum, Cl. tetani.
Corynebacterium diphtheriae, C. equi, C. haemolyticum, C. pseudotuberculosis, C. pyogenes, C. renale.
Edwarsiella tarda.
Erysipelothrix insidiosa.
Escherichia coli, all enteropathogenic serotypes.
Francisella (Pasteurella) Tularensis.
Haemophilus ducreyi, H. influenzae.
Klebsiella—all species and all serotypes.
Legionella—all species and all Legionella-like organisms.
Leptospira interrogans—all serovars.
Listeria—all species.
Mimae polymorpha.
Moraxella—all species.
Mycobacterium—all species.
Mycoplasma—all species.
Neisseria gonorrhoeae, N. meningitidis.
Nocardia asteroides.
Pasteurella—all species.
Plesiomonas shigelloides.
Proteus—all species.
Pseudomonas mallei.
Pseudomonas pseudomallei.
Salmonella—all species and all serotypes.
Shigella—all species and all serotypes.
Sphaerophorus necrophorus.
Staphylococcus aureus.
Streptobacillus moniliformis.
Streptococcus pneumoniae.
Streptococcus pyogenes.
Treponema careteum, T. pallidum, and T. pertenue.
Vibrio cholerae, V. parahemolyticus.
Yersinia (Pasteurella) pestis, Y. enterocolitica.

Fungal Agents

Blastomyces dermatitidis.
Coccidioides immitis.
Cryptococcus neoformans.
Histoplasma capsulatum.
Paracoccidioides brasiliensis.

Viral and Rickettsial Agents

Adenoviruses—human—all types.
Arboviruses—all types.
Coxiella burnetii.
Coxsackie A and B viruses—all types.
Creutzfeldt—Jacob agent
Cytomegaloviruses.
Dengue viruses—all types.
Ebola virus.
Echoviruses—all types.
Encephalomyocarditis virus.
Hemorrhagic fever agents including, but not limited to, Crimean hemorrhagic fever (Congo), Junin, Machupo viruses, and Korean hemorrhagic fever viruses.
Hepatitis associated materials (hepatitis A, hepatitis B, hepatitis nonA-nonB).
Herpesvirus—all members.
Infectious bronchitis-like virus.
Influenza viruses—all types.
Kuru agent.
Lassa virus.
Lymphocytic choriomeningitis virus.
Marburg virus.

Measles virus.
Mumps virus.
Parainfluenza viruses—all types.
Polioviruses—all types.
Poxviruses—all members.
Rabies virus—all strains.
Reoviruses—all types.
Respiratory syncytial virus.
Rhinoviruses—all types.
Rickettsia—all species.
Rochalimaea quintana.
Rotaviruses—all types.
Rubella virus.
Simian virus 40.
Tick-borne encephalitis virus complex, including Russian spring-summer encephalitis, Kyasanur forest disease, Omsk hemorrhagic fever, and Central European encephalitis viruses.
Vaccinia virus.
Varicella virus.
Variola major and Variola minor viruses.
Vesicular stomatis viruses—all types.
White pox viruses.
Yellow fever virus.[2]

(a) *Volume not exceeding 50 ml.* Material shall be placed in a securely closed, watertight container (primary container (test tube, vial, etc.)) which shall be enclosed in a second, durable watertight container (secondary container). Several primary containers may be enclosed in a single secondary container, if the total volume of all the primary containers so enclosed does not exceed 50 ml. The space at the top, bottom, and sides between the primary and secondary containers shall contain sufficient nonparticulate absorbent material (e.g., paper towel) to absorb the entire contents of the primary container(s) in case of breakage or leakage. Each set of primary and secondary containers shall then be enclosed in an outer shipping container constructed of corrugated fiberboard, cardboard, wood, or other material of equivalent strength.

(b) *Volume greater than 50 ml.* Packaging of material in volumes of 50 ml. or more shall comply with requirements specified in paragraph (a) of this section. In addition, a shock absorbent material, in volume at least equal to that of the absorbent material between the primary and secondary containers, shall be placed at the top, bottom, and sides between the secondary container and the outer shipping container. Single primary containers shall not contain more than 1,000 ml of material. However, two or more primary containers whose combined volumes do not exceed 1,000 ml may be placed in a single, secondary container. The maximum amount of etiologic agent which may be enclosed within a single outer shipping container shall not exceed 4,000 ml.

(c) *Dry ice.* If dry ice is used as a refrigerant, it must be placed outside the secondary container(s). If dry ice is used between the secondary container and the outer shipping container, the shock absorbent material shall be placed so that the secondary container does not become loose inside the outer shipping container as the dry ice sublimates.

(d)(1) The outer shipping container of all materials containing etiologic agents transported in interstate traffic must bear a label as illustrated and described below:

ETIOLOGIC AGENTS
BIOMEDICAL MATERIAL
IN CASE OF DAMAGE OR LEAKAGE
NOTIFY DIRECTOR CDC
ATLANTA, GEORGIA
404/633-5313

(2) The color of material on which the label is printed must be white, the symbol red, and the printing in red or white as illustrated.

(3) The label must be a rectangle measuring 51 millimeters (mm) (2 inches) high by 102.5 mm (4 inches) long.

(4) The red symbol measuring 38 mm (1½ inches) in diameter must be centered in a white square measuring 51 mm (2 inches) on each side.

(5) Type size of the letters of label shall be as follows:

Etiologic agents—10 pt. rev.
Biomedical material—14 pt.
In case of damage or leakage—10 pt. rev.
Notify Director CDC, Atlanta, Georgia—8 pt. rev.
404–633–5313—10 pt. rev.

(e) *Damaged packages.* The carrier shall promptly, upon discovery of evidence of leakage or any other damage to packages bearing an Etiologic Agents/Biomedical Material label, isolate the package and notify the Director, Center for Disease Control, 1600 Clifton Road, NE., Atlanta, GA 30333, by telephone: (404) 633–5313. The carrier shall also notify the sender.

(f) *Registered mail or equivalent system.* Transportation of the following etiologic agents shall be by registered mail or an equivalent system which requires or provides for sending notification of receipt to the sender immediately upon delivery:

Coccidioides immitis.
Ebola virus.
Francisella (Pasteurella) tularensis.
Hemorrhagic fever agents including, but not limited to, Crimean hemorrhagic fever (Congo), Junin, Machupo viruses, and Korean hemorrhagic fever viruses.
Herpesvirus simiae (B virus).
Histoplasma capsulatum.
Lassa virus.
Marburg virus.
Pseudomonas mallei.
Pseudomonas pseudomallei.
Tick-borne encephalitis virus complex including, but not limited to, Russian spring-summer encephalitis, Kyasanur forest disease, Omsk Hemorrhagic fever, and Central European encephalitis viruses.
Variola minor, and Variola major.
Variola major, Variola minor, and Whitepox viruses.
Yersinia (Pasteurella) pestis.[2]

§ 72.4 Notice of delivery; failure to receive.

When notice of delivery of materials known to contain or reasonably believed to contain etiologic agents listed in § 72.3(f) is not received by the sender within 5 days following anticipated delivery of the package, the sender shall notify the Director, Center for Disease Control, 1600 Clifton Road, NE., Atlanta, GA 30333 (telephone (404) 633–5313).

§ 72.5 Requirements; variations.

The Director, Center for Disease Control, may approve variations from the requirements of this section if, upon review and evaluation, it is found that such variations provide protection at least equivalent to that provided by compliance with the requirements specified in this section and such findings are made a matter of official record.

[FR Doc. 80–21757 Filed 7–18–80; 8:45 am]
BILLING CODE 4110–88–M

Effective August 20, 1980

[2] This list may be revised from time to time by Notice published in the Federal Register to identify additional agents which must be packaged in accordance with the requirements contained in this part.

Notice to Carrier Label (CDC)

NOTICE TO CARRIER

This package contains LESS THAN 50 ml OF AN ETIOLOGIC AGENT, N.O.S., is packaged and labeled in accordance with the U. S. Public Health Service Interstate Quarantine Regulations (42 CFR, Section 72.25(c), (1) and (4), and MEETS ALL REQUIREMENTS FOR SHIPMENT BY MAIL AND ON PASSENGER AIRCRAFT.

This shipment is EXEMPTED FROM ATA RESTRICTED ARTICLES TARIFF 6-D (see General Requirements 386(d)(1) and from DOT HAZARDOUS MATERIALS REGULATIONS (see 49 CFR, Section 173, 386(d)(3). SHIPPER'S CERTIFICATES, SHIPPING PAPERS, AND OTHER DOCUMENTATION OR LABELING ARE NOT REQUIRED.

____________________ Date

____________________ Signature of Shipper

CENTER FOR DISEASE CONTROL

ATLANTA, GEORGIA 30333

Appendix F

Teaching Aids and Training Courses

SOURCES OF TEACHING AIDS

1. *State Health Laboratory Directors* (loan, no charge)

   a. Safety Management in the Laboratory
      CDC-76-24, John Forney, Ph.D.
      (35 slides and 22-page handout)

Describes safety regulations and requirements under OSHA and CAP; hazards frequently seen; and responsibilities placed upon management, supervisors, and employees. Develops elements of a safety program.

   b. Controlling Infectious Aerosols: Part I—Precautions in Microbiology, produced by CDC, 1976; on loan for return postage only, 16-mm film or videocassette, color and sound.

Demonstrates common hazards in the microbiology laboratory and shows how they produce infectious aerosols. Discusses ways of becoming infected in the laboratory and the factors determining infection.

   c. Controlling Infectious Aerosols: Part II—Minimizing Equipment-Related Hazards, produced by the CDC, 1976; on loan for return postage, 16-mm film or videocassette, color and sound.

Discusses how to use and maintain equipment in a microbiological laboratory. Demonstrates proper use of aerosol-free blender and centrifuge in preventing contamination. Demonstrates use of negative pressure (Class I) and laminar flow (Class II) biological safety cabinets in reducing exposure to infectious agents.

2. *National Institutes of Health* (on loan)
   Division of Safety
   Bethesda, MD 20892

   a. Assessment of Risk in the Cancer Virus Laboratory

   b. Certification of Class II (Laminar Flow) Biological Safety Cabinets

   c. Effective Use of the Laminar Flow Biological Safety Cabinets

   d. Formaldehyde Decontamination of Laminar Flow Biological Safety Cabinets

   e. Fundamentals for Safe Microbiological Research

   f. Hazard Control in the Animal Laboratory

   g. Selecting a Biological Safety Cabinet

3. *National Safety Council*
   444 N. Michigan Avenue
   Chicago, IL 60611

   Introduction to Biohazards Control
   Stock No. 176.54
   Slide-tape cassette

4. *National Audiovisual Center*
   National Archives and Records Administration
   Customer Services Section CL
   8700 Edgeworth Drive
   Capitol Heights, MD 20743-3701
   Telephone number: 1-800-638-1300
   (credit card orders)

   a. Fundamentals for Safe Microbiological Research—A Series
      5 slide sets with instructor manuals
      1 videocassette, with study guide, 1983
      Division of Safety,
      National Institutes of Health
      Series No. A09753/CL

      i. Host Parasite Relationships (Unit 1)

      Describes the role of pH, temperature, and aerobic and anaerobic environment in microbial growth and metabolism; mechanisms of cell growth division; and the replication processes for viruses.

      81 color slides, silent, 176-page
      instructor manual
      Title No. A09754/CL

      ii. Microbial Ecology (Unit 2)

      Covers the chemical and physical properties of the natural habitats that influence the survival and replication of an organism, as well as why an organism is unable to endure and proliferate in natural ecosystems.

      37 color slides, silent, 56-page
      instructor manual
      Title No. A09755/CL

      iii. Principles of Physical and Chemical Containment (Unit 3)

      Defines contamination, explains the principles of contamination control, and lists three types of contamination problems.

      257 color slides, silent, 280-page
      instructor manual
      Title No. A09756/CL

      iv. Biological Containment for Recombinant DNA Molecules (Unit 4)

      Levels of biological containment, categorizing experiments, and the roles played by investigators and NIH committees. Recombinant DNA techniques as defined by the guidelines.

      6 color slides, silent, 84-page
      instructor manual
      Title No. A09757/CL

      v. Laboratory Skills (Unit 5)

      Aseptic technique; principles and purposes of isolating bacteria from mixed broth cultures onto agar plates, staining techniques, and making dilutions and pour plates from liquid suspensions; and the rules for counting colonies after incubation of pour plates.

      123 color slides, silent, 92-page
      instructor manual
      Title No. A09758/CL

   b. Using the Gravity Displacement Steam Autoclave in the Biomedical Laboratory

      Demonstrates aseptic and safe techniques and procedures for preparing, processing, and handling materials undergoing steam sterilization and decontamination.

      29-minute videocassette, study guide
      3/4-inch video no. A10296/CL
      VHS no. A10372/CL
      Beta 2 no. A10371/CL

TRAINING COURSES

1. Control of Biohazards in the Research Laboratory: A training course for biosafety officers.

A one-week course offered annually (third week in July) consisting of lectures and laboratory exercises. Lectures include an overview of cell biology and biotechnology; hazard potential of infectious agents, recombinant DNA, and oncogenic viruses; dissemination of contaminants; equipment designed for safety; containment concepts: primary and secondary barriers; laboratory design criteria; personal practices; safe handling and housing of laboratory animals; principles of ventilation; radiation safety; decontamination, disinfection, and sterilization; how to plan, organize, develop, and conduct a laboratory safety program; emergency procedures; and federal regulations involving laboratory safety.

For further information contact: Course Co-Director, Office of Safety and Environmental Health, The Johns Hopkins Institutions, 2021 E. Monument Street, Baltimore, MD 21205. Telephone number: 301-955-5918.

2. Biological Safety Cabinet Certification Workshop

A one-week course offered annually to anyone responsible for certifying or coordinating the certification of biological safety cabinets. The course consists of lectures and laboratories covering the construction, operation, decontamination, and testing of biological safety cabinets.

For further information, contact Dr. Melvin First, Harvard School of Public Health, 665 Huntington Avenue, Boston, MA 02115. Telephone number: 617-732-1168.

Appendix G

Regulation and Accreditation

REGULATORY DEFINITIONS

Accreditation

A voluntary process recognized as a measure of quality. It is used by some regulatory agencies as one criterion for granting certification and licensure. Standards for accreditation may be similar or identical to standards for licensure. Accrediting organizations are usually based on peer approval, voluntary quality control, education, and consultation.

Licensure

The process by which an agency or government grants permission to persons or facilities meeting predetermined qualifications, to engage in a given occupation, use a particular title, or perform specified functions.

Certification

The process by which a nongovernment agency or association grants recognition to a person who has met certain predetermined qualifications specified by that agency or association.

Equivalency or Reciprocity

A mechanism for comparing programs and functions so that one may be used in lieu of the other. Often the legal body (e.g., a federal agency) maintains authority and stringency over standards even when another organization is identified as an acceptable or equivalent alternative authority.

Guideline

Suggested operating practice or procedure often broadly written. Use of a guideline may be voluntary and is often established by the private sector to self-govern its activities. When issued by a federal agency, strict adherence to the guideline is not obligatory. However, federal guidelines specify one manner of satisfying legal requirements that will be accepted by the agency for the purpose of establishing compliance.

Inspections

Careful, critical, on-the-scene examinations that determine violations against an accepted or legal standard. Inspections may be for accreditation, licensure, certification, or safety. Depending on the inspection type, it may or may not be announced ahead of time.

Law or Act

A legal requirement established by legislative and executive authority, often written in broad, general language. An agency designated by the law must establish or promulgate the specific regulations in order to define how the act will be implemented.

NOTE: Regulatory definitions are modified from Rose, S.L. 1984. A Regulatory Overview. P. 4 in *Clinical Laboratory Safety*. Philadelphia, Pa.: J.B. Lippincott Co.

Regulation

The specific details by which a law is to be implemented. Regulations established by federal agencies are published in the *Federal Register*. Initially these documents appear as proposals so that interested or affected individuals and groups have the opportunity to comment and participate in the rule-making process. After the comments have been reviewed, and either accepted or rejected, the regulation is published again in the *Federal Register* as the "final rule," and the effective date of implementation is given.

Standards

Specific criteria, to be used unmodified for materials, methods, or practices. Use of a standard may be voluntary or mandatory. If it is put forth by a voluntary-consensus standard-setting group, it is usually voluntary; if promulgated by a government agency, it is mandatory. Regulations issued under federal laws are compiled annually into the *Code of Federal Regulations* (CFR) and arranged by subject into assigned Titles: e.g., Title 9, Animals and Animal Products; Title 10, Energy; Title 21, Food and Drugs; Title 29, Labor; Title 39, Postal Service; Title 40, Protection of the Environment; Title 42, Public Health; and Title 49, Transportation.

REGULATORY AGENCIES

1. Federal Agencies

a. Occupational Health and Safety Administration (OSHA)

The Occupational Safety and Health Act of 1970 created OSHA within the Department of Labor. All biomedical laboratories, and their employees, may be subject to the act. Although federal, state, and municipal laboratories were not specifically mentioned in the act, Executive Order 12196 (1980) made all federal agencies subject to the same requirements. The purpose of the act was to reduce occupational injuries and illnesses and each employer is required to furnish to each of his employees a job and a work environment that is free from recognized hazards that are causing, or are likely to cause, death or serious physical harm.

The relevant sections of the Health and Safety Standards (29 CFR Part 1910) are found within the General Industry Standards. Employers of 11 or more employees must maintain records of all injuries and illnesses as they occur. On-the-job accidents resulting in an employee's death or the hospitalization of five or more employees must be reported to the nearest OSHA office within 48 hours. Another relevant section is the Hazard Communication Standard (the "Right-to-Know" rule), which was revised recently to cover all employees exposed to hazardous chemicals, including those working in university research laboratories.

OSHA has documented its interest in the control of biological hazards by publishing an advanced notice of proposed rulemaking concerned with "Occupational Exposure to Hepatitis B Virus and Human Immunodeficiency Virus" (52 FR 45438, November 27, 1987).

Laboratories covered by the act are subject to inspection by OSHA compliance officers and health officers.

b. National Institute for Occupational Safety and Health (NIOSH)

NIOSH was also created by the Occupational Safety and Health Act of 1970, but was made a component of the Centers for Disease Control within the Public Health Service of the Department of Health and Human Services. NIOSH does not regulate, issue, or enforce safety and health regulations. It has the responsibility to undertake research to eliminate on-the-job hazards to health and safety. When requested by employers or employees, health-hazard evaluations of workplaces are carried out.

c. Environmental Protection Agency (EPA)

The Resource Conservation and Recovery Act of 1976, as amended, requires EPA to develop and evaluate environmentally sound methods for management of hazardous waste, including infectious waste. In May 1986, the EPA published its *Guide for Infectious Waste Management* (National Technical Information Service Publication No. PB86-199130, Springfield, VA 22161). This document is intended to provide guidance to persons responsible for infectious waste management at hospitals, medical laboratories, research laboratories, commercial diagnostic laboratories, animal experimentation units, industrial plants and laboratories, and other facilities that generate infectious waste, such as biotechnology companies. Its appendix contains a tabulation of the state regulations pertaining to infectious waste management, including a summary of the requirements and the identity of the responsible state agency.

d. Food and Drug Administration (FDA)

The FDA issues regulations pertaining to the safety, efficacy, and labeling of drugs and medical devices, including diagnostic reagents and laboratory equipment, under the provisions of the Food, Drug, and Cosmetic Act, as amended, and of biological products under the provisions of the Public Health Service Act. The Office of Medical Devices in the Center for Devices and Radiological Health (CDRH) has the responsibility for laboratory instruments, diagnostic agents, and reagents. The Center for Drug Evaluation and Research (CDER) is responsible for the premarket testing and approval of drug products for human use, and the Center for Biologics Evaluation and Research (CBER) similarly controls biological products for human use, including blood and blood products, bacterial and virus vaccines, and certain diagnostic materials for dermal tests and laboratory tests. The regulations do not directly address issues of safety in biomedical laboratories. These matters are affected indirectly in the regulations covering Current Good Manufacturing Practices (21 CFR, Part 211) and Good Laboratory Practice for Nonclinical Laboratory Studies (21 CFR, Part 58), which apply to laboratories working with infectious materials in the course of research, development, testing, and manufacture of controlled products.

e. National Institutes of Health (NIH)

The National Cancer Institute issues guidelines for the control of contamination in facilities working with cancer viruses. In conjunction with the Centers for Disease Control, the NIH published guidelines for *Biosafety in Microbiological and Biomedical Laboratories*, which has been reproduced as Appendix A in this publication. In addition the NIH publishes guidelines on various other aspects of laboratory safety, as well as on laboratory animal care. The NIH *Guidelines for Research Involving Recombinant DNA Molecules* were first published in 1976. They are updated periodically, most recently in 1986.

f. Centers for Disease Control (CDC)

The CDC has the responsibility for controlling the interstate shipment of etiologic agents under the Public Health Service (PHS) Interstate Quarantine regulation (42 CFR, Section 72.25). The importation into the United States of etiologic agents and vectors of human disease, and subsequent transfer to other laboratories, requires the issuance of a permit to the recipient by the CDC under the provisions of the PHS Foreign Quarantine regulation (42 CFR, Part 71, Section 71.156). In addition to the CDC/NIH biosafety guidelines mentioned above, CDC publishes recommendations for various other aspects of laboratory safety and operations.

g. U.S. Department of Transportation (DOT)

In addition to the CDC, the DOT has responsibility for regulating shipment of etiologic agents, diagnostic specimens, and biological products that are shipped in the United States. Its regulation differs in part from the CDC requirements in that packaging materials must have been proven to be adequate to contain infectious material under a variety of environmental and test conditions, and control is exerted over both intrastate and interstate shipments.

h. U.S. Department of Agriculture (USDA)

The importation of etiologic agents of plant and animal diseases, serum specimens, and other materials of animal origin requires a permit that must be obtained from the appropriate authorities in the USDA. (See Chapter 3, Section G.)

i. U.S. Postal Service (USPS)

The USPS requirements for the mailing of "diseased tissues, blood, serum, and cultures of pathogenic microorganisms" are essentially the same as those of the CDC and DOT.

j. U.S. Nuclear Regulatory Commission (U.S. NRC)

The U.S. NRC strictly regulates the possession, use, and disposal of radioactive materials under a licensing system.

k. Health Care Financing Administration (HCFA)

By overseeing both Medicare and Medicaid programs and related medical care quality control, the HCFA has a direct relationship with hospital laboratory services and indirectly influences safety concerns. HCFA currently has the responsibility for licensing and inspecting clinical laboratories, which are subject to the Clinical Laboratory Improvement Act of 1967.

l. Office of Science and Technology Policy (OSTP)

The *Coordinated Framework for Regulation of Biotechnology: Announcement of Policy and Notice for Public Comment (1986)* describes federal regulatory policy for ensuring the safety of biotechnology research and products. This document emphasizes mostly technologies for emerging genetic manipulation, such as recombinant DNA, and summarizes previously published policies of FDA, EPA, USDA, OSHA, and NIH.

2. State and Local Government Regulations

States vary in their regulations applicable to biomedical laboratories and the handling of biohazardous waste. One summary of state laws regarding hazardous waste is found in the appendix of the Environmental Protection Agency (EPA) *Guide for Infectious Waste Management* (1986). This publication can be obtained from the Superintendent of Documents, U.S. Government Printing Office, Washington, DC 20402-9325.

III. ACCREDITING BODIES

1. College of American Pathologists (CAP)

The College of American Pathologists offers a voluntary accreditation program for clinical laboratories. The requirements for safe work practices include biosafety. The inspector, a practicing pathologist, completes an extensive checklist during a site review of the procedures and facilities. Cited deficiencies must be corrected before accreditation is granted. CAP accreditation can exempt a laboratory from federal inspection for interstate licensure and is recognized by the Joint Commission on Accreditation of Healthcare Organizations.

2. Joint Commission on Accreditation of Healthcare Organizations (JCAHO)

The Joint Commission on Accreditation of Healthcare Organizations has a number of safety requirements similar to those of the CAP cited above. In addition, it has extensive safety requirements for other areas of the hospital.

3. American Association for Accreditation of Laboratory Animal Care (AAALAC)

One of the considerations for certification by the Council of Accreditation of AAALAC is indirectly related to control of biohazards encountered during the course of experiments involving laboratory animals. The certification process involves an extensive on-site inspection, and recertification is a biennial event. Recognition by AAALAC attests that the recipient of accreditation is in compliance with the Laboratory Animal Welfare Regulations (9 CFR, Subchapter A, Parts 1, 2, and 3) and adheres to the standards set forth in the *Guide for the Care and Use of Laboratory Animals*, DHHS Publication No. (NIH) 86-23 (revised 1985). Both of these publications can be obtained from the Superintendent of Documents, U.S. Government Printing Office, Washington, DC 20402-9325.

4. State Health Departments

Health departments of various states are required to perform inspections of medical laboratories that receive payment from Medicaid. They have safety requirements similar to those of CAP and JCAHO cited above. Medicaid and the state health departments will look for safety deficiencies similar to those described for the JCAHO and CAP programs.

Appendix H

List of Abbreviations

AIDS	acquired immunodeficiency syndrome
AOMA	American Occupational Medical Association
APHIS	Animal and Plant Health Inspection Service
ATCC	American Type Culture Collection
BTU	British thermal unit
CDC	Centers for Disease Control
CJA	Creutzfeldt-Jakob agent
CJD	Creutzfeldt-Jakob disease
DNA	deoxyribonucleic acid
EPA	Environmental Protection Agency
HAV	hepatitis A virus
HBV	hepatitis B virus
HEPA	high-efficiency particulate air
HIV	human immunodeficiency virus
IND	investigational new drug
MSDS	material safety data sheet
NIH	National Institutes of Health
NRC	Nuclear Regulatory Commission
OSHA	Occupational Safety and Health Administration
PEE	postemployment evaluation
PME	periodic monitoring examination
PPE	preplacement examination
RCRA	Resource Conservation and Recovery Act
RSO	radiation safety officer
SOP	standard operating procedure
USAMRIID	U.S. Army Medical Research Institute for Infectious Diseases
USDA	U.S. Department of Agriculture
USPHS	U.S. Public Health Service
VS	Veterinary Services

Index

C

D

E

K

L

M

N

O

P

Q

R

S

V

W

Y

Z